STARSEEKERS

—COLIN WILSON—

Book Club Associates
London

CONTENTS

PART I

ANCIENT
COSMOLOGY

CHAPTER ONE

THE GREAT STONE OBSERVATORIES

At the age of thirty-eight, worn out by alcohol, drugs and failure, Edgar Allan Poe decided to make one last frantic bid for success. He would write a book about the creation and destruction of the universe.

It is a pity that Poe has left us no account of how the revelation came to him. For it was, undoubtedly, a revelation – one of those lightning flashes of vision that convince the seer that he has seen into 'the mind of God'. Poe's first biographer says: 'no other author ever flung such an intensity of feeling, or ever believed more steadfastly in the truth of his work, than did Edgar Allan Poe in this attempted unriddling of the secret of the universe.'[1] Poe's title tells us as much: *Eureka*, 'I've found it!', the legendary cry of Archimedes when he discovered the law of floating bodies and leapt out of his bath to run naked through the streets of Syracuse.

Poe composed the book in a fever of inspiration during the winter of 1847. He was, says his biographer, 'a man exalted'. He persuaded a publisher of the importance of the book, and advised him to print a first edition of fifty thousand copies. He hired a lecture hall in New York, and left his audience excited and impressed by his vision of creation and destruction. He was convinced not only that *Eureka* would make his fortune, but that it would establish him as one of the greatest thinkers of the age.

Eureka appeared in March 1848, in an edition of five hundred copies, and it was a flop. The few critics who noticed it were contemptuous. A typical comment came from the anonymous reviewer of the *Literary World*, who said that Poe's claims to understand the origin of the universe 'must be set down as mere bold assertion, without a particle of evidence. In other words, we should term it *arrant fudge*.'[2] Poe wrote a furious letter to the editor; his indignant howl of anguish would be funny if it were not so obviously the cry of a soul in torment. But one phrase reveals his continuing conviction of the book's greatness. Defending himself against the accusation that his main ideas were stolen from Laplace, Poe declares: 'The ground covered by the great French astronomer compares with that covered by my theory, as a bubble compares with the ocean on which it floats.'[3] Pierre Simon de Laplace, it may be mentioned in passing, was the author of the most important volume on celestial mechanics since Newton's *Principia*.

And is *Eureka* Poe's masterpiece, as at least one critic believes?[4] The answer, I am afraid, has to be 'no'. This portentous attempt to describe the 'physical, metaphysical, mathematical, material and spiritual universe' is abstract and pretentious. Moreover, some imp of the perverse drove him to introduce it with an epistle in his silliest 'humorous' manner, attacking various predecessors in cosmology, 'a Turkish philosopher called Aries and surnamed Tottle', 'whose illustrious disciples were one Tuclid, a geometrician, and one Kant, a Dutchman'. After this buffoonery, Poe goes to the opposite extreme, and talks darkly about ratiocination, axiomatic suppositions and 'the impossibility of attributing supererogation to Omnipotence'. As if to convince the reader that these pseudo-profundities deserve the closest attention, he places every other word in italics.

How do we account for this aberration in a man of genius? Megalomania? Alcoholic insanity? Neither, I believe, is the correct solution. For here we come to the most surprising part: anyone who persists and makes his way through this jungle of verbiage will discover that *Eureka* contains some staggering, almost incredible insights. To understand how remarkable they were, we have to try to forget the scientific advances of the past century and a half, and imagine ourselves in America in the 1840s. The atomic theory of John Dalton was less than half a century old, and atoms were still thought of as tiny hard balls; yet Poe asserts confidently that matter can be reduced to attraction and repulsion – an observation that would only be confirmed with Ernest Rutherford's discovery of the structure of the atom fifty years later. The great cosmologist Laplace, whose influence is obvious throughout *Eureka*, set out to demonstrate that the solar system is as stable as a grandfather clock, while Alexander von Humboldt's *Kosmos*, the other major influence on Poe's thinking, obviously takes the same view of the entire universe. Yet Poe declares that the universe began as a single ball of matter, and then exploded outwards to form the stars – anticipating by seventy years Willem de Sitter's theory of the expanding universe, first published in 1917. Poe also asserts that the universe will eventually collapse inwards again, and end in annihilation – and here he has outlined a theory of the universe that we owe to modern radio astronomy. According to this theory the universe began with a 'big bang', will eventually reach a certain limit – because the speed of expansion is less than the speed of 'escape' – and will then contract again, ending as a collapsing star or black hole. Poe also throws off the casual suggestion that space and time are the same thing, an insight that seemed obvious nonsense at the time, and that did not begin to make sense until Einstein's appearance. Poe even recognized that the Milky Way is a galaxy, and that galaxies are island universes, and not mere star clusters – another notion that would not be confirmed until the twentieth century. (I have a compendium on astronomy published in about 1930, and a photograph of the Milky Way has a caption that includes the words: 'some astronomers are of the opinion that it marks the limit of the visible universe, and that an immense space, empty to our senses, lies on the further side of it.') And when Poe states that the universe ends in annihilation, and then begins all over again, he anticipates one of the most recent theories of cosmology: that a black hole does not continue to collapse indefinitely, but that it finally reaches a limit, and then explodes again.

'What I propound here is true,' Poe insists in the preface, and he repeats this conviction with megalomaniac obstinacy throughout the book. In fact, he continued to believe it throughout the year of life that still remained to him (he died after a drunken debauch in Baltimore in October 1849). It was natural for most of Poe's biographers to assume that this was an attempt at self-deception – the desperation of a man of genius whose work was ignored; but then, they were unaware of the basic accuracy of his insights. The more we consider these insights, in the light of what we know about modern cosmology, the more it begins to look as if Poe was in the grip of some powerful intuition that carried him far beyond the speculations of Humboldt and Laplace.

It would not be the first time in history this has happened. The Greek philosopher Heraclitus (*c.* 540–480 BC) stated that the universe is a living organism that is subject to birth and death; when it dies, it leaves behind a seed, from which it grows again. Everything derives from fire, and moves in a cyclical process. Again, we seem to have a vision of an expanding and contracting universe.

Poe himself would have scorned the idea that knowledge can be obtained through direct intuition. Although he insists that the basic ideas of *Eureka* are a matter of intuition – and that this is why he is so sure they are true – he then hedges his bets by giving a thoroughly rationalistic definition of intuition: '. . . the conviction arising from these inductions or deductions of which the processes are so shadowy as to escape our consciousness, elude our reason, or defy our capacity for expression.'[5] In other words, intuition is subconscious observation. Yet his final appeal to a Creator – to set the whole operation in motion – suggests that he is less of a rationalist than he sounds. Certainly, Heraclitus would have accepted a quite different definition – the notion of intuition as a kind of direct insight that defies rational processes.

Edgar Allan Poe.
His 'crank' theories of the universe have
been borne out by modern science

Now this notion is, in fact, less preposterous than it sounds. The branch of science known as split-brain research tells us that the cerebral hemispheres of the brain have quite different functions. The left is the rationalist, the logician – the scientist; the right is the intuitionalist, the apprehender of patterns – the artist. Odder still, the person you call 'you' lives in the left half. If the knot of nerve fibre joining the two halves is severed – as it may be to prevent epilepsy – the patient virtually turns into two different persons. If he is shown an apple with the left half of the brain (which is connected to the right visual field) and an orange with the right, and he is asked, 'What have I just shown you?', he replies, 'An apple.' Asked to write what he has seen with his left hand (connected to the right hemisphere), he writes, 'An orange.' If he is not allowed to see what he has written, and he is asked what it is, he replies, 'Apple.'

The 'left you' is the one who uses language; the 'right you' is non-verbal. Yet it *can* convey information. In an experiment performed by the psychologist Roger W. Sperry, red or green lights were flashed at random in the left visual field of a split-brain patient, and he was asked to guess what colour had just been flashed. The score should have been random, because the left-brain ego does not see the colours. In fact, the score was well above chance. The patient would say, 'Red', and then jump, as if someone had nudged him in the ribs, and say, 'No, sorry, green.' The right brain had heard the incorrect answer, and conveyed the information by, so to speak, kicking him under the table. The left and right brains behave like two different entities; in fact, like two independent minds.

The significance of all this will become clear in a moment, when we discuss the very early history of astronomy. Meanwhile, it is important to grasp the full extent of the difference between these Siamese twins and their ways of apprehending the universe.

The left brain could be compared to a microscope; it is concerned with detail, with particularities. The right is more like a telescope; it is more concerned with emerging patterns. This means, for example, that ordinary calculation – like adding up the grocery bill – is performed by the left, while 'creative mathematics' – let us say, working out some interesting geometrical problem – involves the right.

This enables us to consider a concrete example of 'direct intuition'. The seventeenth-century mathematician Pierre de Fermat believed that a certain formula* always generates prime numbers – that is, a number that cannot be divided by any other (for example 3, 5 and 7). The first five numbers generated by this formula are 3, 5, 17, 257 and 65 537 – all known to be primes. The sixth number in the series – in which $n = 5$ – is enormous: 4 294 967 297. Is this number a prime?

Now oddly enough, there is no general method for discovering whether a given number is a prime, except by painfully dividing every other number into it – a process of elimination. Yet in the mid-nineteenth century a Canadian 'lightning calculator', a boy named Zerah Colburn, was asked whether this number is a prime and replied after a moment's thought, 'No, it can be divided by 641.' He could not explain how he arrived at this answer. But it cannot have been by any 'rational' process since (apart from the elimination method) none exists.

The whole topic of calculating prodigies is still something of a mystery. They are usually uneducated, and not particularly intelligent – some have literally been idiots. Their unusual powers tend to vanish as they get older, as if the coming of adulthood, with its practical problems, demanded a rearrangement of their mental powers. And

* The formula is $2^{2^n} + 1$.

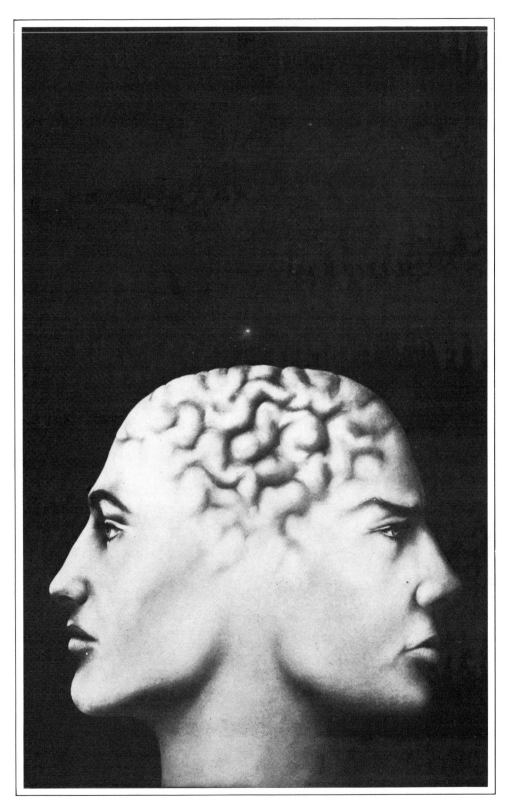

13

since it is the left brain that copes with down-to-earth problems, this suggests that the powers of lightning calculation are associated with the right – a conclusion that seems to be supported by Colburn's feat.

Robert Graves has an interesting anecdote of a calculating prodigy in his short story *The Abominable Mr Gunn*. Grave's form-master – Mr Gunn – had set the class a difficult arithmetical problem. One boy named Smilley wrote the answer immediately in his exercise book; when Mr Gunn wanted to see his calculations, he explained that 'it had just come to him'. Mr Gunn suggested that Smilley had looked up the answer in the back of the book; Smilley admitted that he had – afterwards – and added that the last two figures given in the answer were wrong anyway. Mr Gunn declined to believe him, and sent him off to be caned.

In the same story, Graves describes a related experience of his own: 'One fine summer evening as I sat alone on the roller behind the cricket pavilion . . . I received a sudden celestial illumination: it occurred to me that I knew everything. I remember letting my mind range rapidly over all its familiar subjects of knowledge; only to find that this was no foolish fancy. I did know everything.' He goes on to qualify this: 'To be plain: though conscious of having come less than a third of the way along the path of formal education . . . I nevertheless held the key to truth in my hand, and could use it to unlock any door. Mine was no religious or philosophical theory, but a simple method of looking sideways at disorderly facts so as to make perfect sense of them.'[6] He tells how, the following day, he tried to record his 'secret' in a notebook, but his mind ran too fast for his pen (a typical problem associated with right-brain insights), and he began crossing out – a fatal mistake. When he tried again that evening, the insight had vanished.

Graves has also recorded how, many years later, another 'unsolicited enlightenment' caused his mind to outrun his pen. He was reading the *Mabinogion*, 'when I suddenly knew (don't ask me how) that the lines of the [Taliesin] poem, which had always been dismissed as deliberate nonsense, formed a series of early mediaeval riddles, and that I knew the answer to them all – although I was neither a Welsh scholar, nor a mediaevalist. . . .'[7] These insights, and others like them, formed the basis of *The White Goddess* – a book that argues that such flashes of knowledge belong to the realm of the moon goddess. In fact, the basic argument of *The White Goddess* is that there are two distinct forms of knowledge: solar knowledge – that rational, scientific knowledge which has become typical of Western man – and 'lunar' know-ledge – which springs from intuition, from the unconscious. Lunar knowledge, which is connected with poetry and magic and the irrational, is far older than solar know-ledge, which finally ousted it.

The distinction may sound artificial. After all, knowledge in itself is not really 'lunar' or 'solar': only the *method* by which I arrive at it can be labelled intellectual or intuitive. But then what Graves is trying to point out is that Western man now makes an unconscious assumption that knowledge is intellectual by nature, and that intuition is simply a crude way of groping towards intellectual conclusions. In other words, anything that can be seen by moonlight can be seen better still by sunlight. This is untrue; certain things – like candle flames – become invisible in strong sunlight. And the human spirit itself resembles a candle flame in this sense. It can be most clearly seen – or felt – by a magical half-light. Moreover, when it is seen and felt, man's relation to his universe is altered. He becomes aware of himself as an *active* principle, rather than as a passive object. While he accepts himself as passive – as a creature who

The Babylonian world system, based on the much earlier Sumerian cosmology

is acted upon – his relationship to objects (his knowledge) is somehow false. Thus our 'solar' view of knowledge falsifies the universe.

And so, by a rather circuitous route, we come back to cosmology – that is, man's relation to the universe. And we find that Graves's view of knowledge is the key to an interesting mystery.

The 'first astronomers', in the modern sense of the word, were the Babylonians or 'Chaldeans', who merged with the Assyrians – being conquered by them – in the seventh century BC. In the Bible, the word Chaldean is usually synonymous with 'wise man' – meaning a magician or astrologer. Astronomy, in the sense of scientific observation of the heavens, begins (roughly) around 800 BC. (The Chinese had their own tradition, but it will not concern us in this book.) Yet the Mesopotamian peoples had been studying the heavens for at least three thousand years by this time, and noting the movements of the moon. The reason, obviously, is that any society that depends on agriculture needs to be able to predict the seasons, and the moon forms a highly convenient clock.

As for more northerly countries of Europe – like England and France – it was assumed that their knowledge of the heavens consisted of little more than this lunar rule of thumb. Their natives were too concerned with fighting, and with conducting rituals in temples like Stonehenge and Carnac, to be interested in the heavens.

In 1897, a disputatious note was sounded when a Frenchman named Félix Gaillard published *L'Astronomie préhistorique*, in which he argued that prehistoric stone monuments – like the great avenues of stone at Carnac – were each a kind of astronomical calculator, built to determine such matters as solstices and the nineteen-year lunar

cycle. But Gaillard's arguments were less rigorous than they might have been – a point made by the British astronomer Sir Norman Lockyer who, in 1909, published a book on Stonehenge in which he argued that this monument was also an astronomical calculator. Lockyer had become interested in the astronomical alignments of the Greek temple at Eleusis as early as 1890, and in 1901 he began to study Stonehenge, as well as other English stone circles like the Merry Maidens in Cornwall. He concluded that Stonehenge was virtually a stone calendar, a giant sundial – whose purpose, presumably, was to tell the Neolithic farmers when to sow their crops. Oddly enough, Lockyer failed to see that he could make an even more convincing argument for Stonehenge as a lunar calendar.

In 1934 a Scots engineer named Alexander Thom visited the stone circle at Callanish in the Hebrides, and realized that the avenue of menhirs was pointing at the pole star, and ran from north to south. But in prehistoric times, the pole star was not in its present position in relation to the earth; so the men who constructed Callanish must have had to align the stones on true north without its help, which argued considerable engineering ability for Neolithic or Bronze Age farmers. Thom began to study stone circles, and became convinced that most of them were basically astronomical calculators, intended to compute the movements of the moon. He observed that many 'circles' were irregular, containing 'bulges' – looking as if the builders were simply incompetent. But close examination of these irregular circles convinced him that they were designed with great geometrical precision, and that the aim was to make the circumference precisely three times the diameter, instead of $3.142 - \pi$.

In 1963, Gerald Hawkins, a professor of astronomy at Boston, caused considerable excitement by publishing an article in *Nature* supporting – and greatly elaborating on – Lockyer's ideas. One of the chief problems in deciding whether Stonehenge could have been used as a 'calendar' is that there are literally thousands of possibilities, and these would have to be worked out in terms of the position of the heavenly bodies at the date of the construction of the monument. Such calculations would take a mathematician many years. But a modern computer can do it in seconds. Hawkins fed his complex data into a computer, and was astounded to find the result quite unambiguous: Stonehenge was a precise solar *and* lunar calendar. Hawkins went on to write a book called *Stonehenge Decoded*, which became an instant best seller – although many readers must have found its revelations obscure and disappointing. At this point, the British astronomer Fred Hoyle entered the controversy, and in a paper published in *Nature* in 1966 advanced his own theory of how the outermost circle of Stonehenge with its 'Aubrey Holes' (named after the antiquarian John Aubrey) could have been used to compute the moon's 18.61-year cycle around the earth's ecliptic (the apparent path of the sun around the earth). Hoyle was obviously in basic agreement with Hawkins about the purpose of Stonehenge, and his support caused something of a sensation. For it must be realized that Lockyer's views on Stonehenge had never achieved wide acceptance in the scientific community. In a standard work on Stonehenge published in 1956, R. J. C. Atkinson had no doubt about the 'unsoundness of the theory'; on the subject of the purpose of Stonehenge, he stated emphatically, 'We do not know, and shall probably never know.'[8] The few archeologists to whom Thom's views were known – through publication in technical journals – were

Opposite above: Stone circle at Callanish on the Isle of Lewis
Opposite below: The stone alignments of Carnac in Brittany

inclined to regard him as a crank (as many still do). Atkinson himself dismissed Hawkins's work as 'unconvincing, tendentious and slipshod', and attacked him in an article entitled 'Moonshine on Stonehenge'. So this acceptance of Hawkins – and Lockyer – by a major British astronomer was something of a revolution; a 'crank theory' had become almost respectable.

Perhaps Atkinson can hardly be blamed for dismissing Lockyer's astronomical theory. When he discusses the dating of Stonehenge, he mentions that radio-carbon dating of charcoal found in one of the Aubrey Holes gives a date of 1848 BC, plus or minus 275 years. Thus its oldest date would be about 2100 BC. As for the builders of this outermost bank and ditch, Atkinson believes that they were 'Secondary Neolithic' people. This needs some explanation. Before the building of Stonehenge, according to Atkinson, the native inhabitants of the region were savage nomads who lived by hunting and fishing. Then in around 2300 BC colonists began to arrive from Europe, bringing with them the basic necessities for farming – horned cattle and grain. The local savages were influenced by these colonists – known as the Windmill Hill people – and began themselves to keep herds of cattle. It was these nomadic herdsmen, whom Atkinson compares to modern tinkers and gypsies, who were the 'Secondary Neolithic' people who built the first part of Stonehenge – the ditch and Aubrey Holes.

Clearly, you would not expect much sophistication or culture from primitive tinkers and gypsies. And this was one reason why Atkinson found Hawkins's theory preposterous.

Yet new radio-carbon dating was to reveal that the Aubrey Holes were far older than Atkinson supposed. Carbon dating depends on a radioactive substance known as Carbon 14, which is absorbed from the atmosphere by all living creatures. When they die, the carbon decays into nitrogen; and as it decays at a fixed rate, a measure of the Carbon 14 in a bone or a piece of wood can give an accurate notion of its age.

The carbon dating mentioned in Atkinson's book took place in the early 1950s, and it was based on the assumption that the amount of Carbon 14 in the earth's atmosphere is more or less constant. Then someone had the idea of checking this by examining the rings of Bristlecone Pines in California, some of them 4000 years old. And these revealed that the amount of Carbon 14 in the atmosphere fluctuates considerably. Adjustments had to be made to all dates of more than a few hundred years. Suddenly, the charcoal in the Aubrey Holes, which Atkinson had dated somewhere between 1700 and 1900 BC, was revealed to be about a thousand years older.

This changes our whole concept of 'prehistoric' society. 1700 BC was Early Bronze Age – basically the age that Homer writes about in *The Iliad*. 2900 BC is very definitely Neolithic – the New Stone Age, a period when, as Atkinson said, Britain should have been populated by savages. But would nomadic savages have known enough about astronomy to build a sophisticated calculator like Stonehenge 1? Clearly not. The latest view has been stated by Professor Euan MacKie in *Science and Society in Prehistoric Britain*. According to MacKie, the people who built Stonehenge 1 (and the nearby Silbury Hill, the largest man-made mound in Europe) were Neolithic farmers who belonged to a highly stratified 'theocratic' society – a society ruled by priest-kings (or at least, one in which priests played a dominant role). These priests were 'a professional class of wise men' who could organize enormous numbers of men,

Opposite above: Detail of a geometrical ground plan of Stonehenge by William Stukely, the English antiquary sometimes known as 'The Arch Druid', 1723

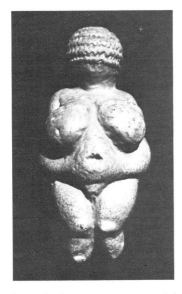

and considerable resources, to construct these vast and time-consuming works. In fact, the British equivalent of 'Chaldeans'.

The idea sounds astonishing and, at first sight, thoroughly far-fetched. But MacKie supports it with a great deal of careful analysis. He attaches particular importance to the discoveries made between 1967 and 1971 at various Henge sites in southern England. These revealed that the larger Henges (ritual sites – the word derives from Stonehenge) contained massive wooden-roofed buildings. But the domestic refuse found at these sites did not reveal the things you might expect if the inhabitants were an ordinary Neolithic tribe – for example, although there were pig bones, there were no pig skulls. This argues that the pork was brought to the site from elsewhere, ready for cooking. Which in turn suggests that the inhabitants at the site – presumably priests, since Henges are 'temples' – were a special class, supported by the surrounding population.

Which leaves the obvious question: why were primitive farmers so interested in the phases of the moon? They had no need to make elaborate calculations to plant their simple crops; even the study of the moon was unnecessary, since the most inexperienced farmer would realize that spring wheat should be planted when the first signs of spring arrive.

Besides, even assuming that the Windmill Hill people arrived several centuries before the date assumed by Atkinson (2300 BC), the Henges still represent a considerable mystery. Elaborate religions do not spring into existence overnight, or even in a single generation (unless there is a 'messiah' to aid the process). If the Henges were built by hunting peoples who had learned a little about farming from the Windmill Hill colonists, we are probably justified in assuming that their interest in the sun and the moon dated from a period before they learned about farming.

And there is, in fact, an interesting, if controversial, piece of evidence to show that this is so. In the 1960s, a science writer named Alexander Marshack came across a photograph of a piece of bone dating from about 6000 BC; it contained a number of curious indentations that resembled some kind of writing. But writing was not invented until around 3500 BC by the Sumerians. Marshack began a four-year course of study of hundreds of similar pieces of incised bone, some dating back thirty-four thousand years, when Europe was inhabited by Cro-Magnon man, the direct ancestor of present-day man. Marshack concluded that the 167 marks on the eight-thousand-year-old bone were a record of phases of the moon over a six-month period. A plaque of reindeer antler found in the Dordogne confirmed this view. It contained a winding row of sixty-nine small holes, and microscopic examination showed that they had been incised with different tools at different times – suggesting that this was more than a decoration. Marshack has made a convincing case that each of the dots represents a different night, and that the snakey path is an attempt to represent the moon's rising and setting places. The similarity of markings on the most ancient artefacts suggested that Cro-Magnon man was also interested in the phases of the moon.

Why should this be so? Here we are on firmer ground. The most primitive religion known to history is that of the mother goddess. Female figurines – sometimes called Venus figurines – date back to the Aurignacian (Cro-Magnon) period, more than twenty thousand years ago. Jacquetta Hawkes writes: 'Their most characteristic products were female figurines in bone, ivory or stone, showing enormous breasts, bellies, buttocks and thighs. . . . Essentially . . . they came from an inner vision of fertility and motherhood. In this sense, they can be said to be the first evolved religious symbols. . . . The continuity of meaning and emotion between these works and the 'Mother Goddess' figurines of the New Stone Age . . . can hardly be denied.'[9]

There is significance in the fact that they should also be called Venus figurines; for in many ancient mythologies, the earth goddess often became oddly confused with the moon goddess and with Venus. So the Egyptian Isis, the mother goddess, is also identified with the Greek Demeter, the harvest goddess, and with Selene, the moon goddess, as well as with Aphrodite (the Greek name for Venus). The Semitic names for the mother goddess are: Inanna, the Sumerian mother goddess; Ishtar, the

Aphrodite, Greek goddess of love and wedlock; by Praxiteles, fourth century BC
Opposite: The Willendorf Venus. This figurine from Austria dates from 2500 BC

Babylonian mother goddess (also Venus); and Astarte, the Canaanite earth goddess, also identified with Venus and the moon. If this seems confusing, we must bear in mind that most of these primitive religions had a primary god and goddess – like Isis and Osiris – and that a primitive mysticism led to an identification of the leading goddesses within the pantheon. Venus, the morning and evening star, was one of the most prominent bodies in the heavens after the sun and moon.

We can understand, then, why ancient man should have built temples to the sun and moon: he worshipped both.

But why should such temples have also been calculators? Of course, on one level the answer is obvious. The moon encircles the earth every $29\frac{1}{2}$ days. The earth encircles the sun every $365\frac{1}{4}$ days – which is why we have a leap year every four years. So the year

Women under the influence of the moon. A seventeenth-century French engraving

contains 12 months and 11 odd days. The Jews managed to 'balance' the calendar by using a 19-year period of 235 months, in which seven of the years contained 13 months. If Hawkins and Hoyle are correct, the builders of Stonehenge worked out a complex method involving a 56-year period – three times the moon's 18.61-year cycle round the ecliptic. But *why* did they want this kind of accuracy? What did it matter to such primitive people?

The answer, quite clearly, is that it was connected with their religious worship. The sun and moon were gods; therefore it was incumbent on the priests to be able to forecast their movements *precisely*. However, that still leaves unanswered the question: why did ancient man feel such religious awe in the face of the earth, the sun and the moon – not to mention Venus? To feel it about thunder, lightning, hurricanes – that is comprehensible. But as far as we know, Neolithic man set up no elaborate temples to thunder and lightning – certainly nothing to rival Stonehenge or Avebury.

And it is at this point that this argument needs to take a 'leap in the dark'. For me, the most significant question is how ancient man could allow the earth goddess to overlap with the moon goddess and Venus. There seems to be no similar tendency to allow the gods to overlap – for example, to allow the Egyptian Ra, the sun god, to blend with Osiris, the earth god, or Horus, the sky god, or the Greek Jove to blend with Mars or Apollo. But then, the moon differs from all other heavenly bodies in one basic respect: it has a direct and obvious influence on the earth – not just on the tides, but on the minds of human beings. Arnold M. Lieber, a member of the clinical faculty at the University of Miami School of Medicine, has written an important study of 'the lunar effect' on human emotions,[10] and he has no doubt that his own patients are influenced by the moon. He quotes policemen, magistrates and ambulance men who have observed the rise in the crime and accident rate at the full moon. Most doctors – including my own local GP – would agree with him. Most country areas have certain

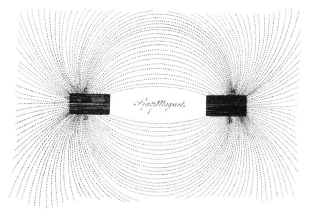

Lines of magnetic force. An eighteenth-century demonstration using magnets and iron filings

characters who, if not 'lunatics', are undoubtedly inclined to become a little odd at the time of the full moon.

Lieber noted something else – that his patients could be equally influenced by sunspots. Here again there is much scientific evidence – for example, in the researches of Harold Saxton Burr and F. S. C. Northrop, which showed that sunspots affected the 'life fields' of plants and animals. Burr and Northrop attached sensitive voltmeters to trees, and discovered that electrical fields were associated with their vital processes, and that these fields varied with day and night, with the seasons, with the cycles of the moon, and with sunspot activity. Attached to animals or human beings, the volt-meters could detect emotional stress, ovulation and incipient – as well as actual – illness. Their paper 'The Electrodynamic Theory of Life' suggested that living cells are held together by these electrical fields and shaped by them – rather as iron filings are shaped by the field of a magnet.

In a book called *The Climatic Threat*, the scientist John Gribbin documents the influence of the eleven-year sunspot cycle on the weather, and therefore on crops and grazing animals. (He mentions, for example, that 'in the early 1970s, farmers in England were increasing their stocks of cattle after the successes of previous years, just at a time, we now see, when the solar cycle effect was working against them – with disastrous results'.)[11] And in a book called *The Jupiter Effect*, he remarks: 'The gravity of the planets can affect the sun, through tidal interactions, and the disturbances of the sun can influence the earth through changes in the magnetic field which links all the planets in the solar system.'[12] He adds that on rare occasions, these small effects can add up to produce dramatic results – like earthquakes – and that the next such alignment of the planets is due in 1982.*

· Just as important as the effects on the earth's field are the effects on the human body. In 1938, a Japanese doctor named Maki Takata noticed that a test he had developed – for measuring the rate at which albumen curdles in the blood – was going haywire. The 'flocculation index', as this curdling is called, had risen quite abruptly; moreover, investigation showed this was happening all over Japan. Takata discovered that this was the result of unusual sunspot activity, and it also led him to notice that the flocculation index jumped just before dawn, as the blood responds to the rising sun (even though it is still below the horizon). Takata found himself speculating whether the index would drop during eclipses, when the moon placed itself between the earth and the sun. In 1941, 1943 and 1948, he was able to test subjects during total eclipses, and found that his hypothesis was correct.

Dr Leonard J. Ravitz, of the Virginia Department of Health and Education, investigated the influence of moon cycles on mental patients in the late 1950s. He discovered that there is a difference in electrical potential between the head and chest, and that in mental patients this difference is greater than in normal people. This difference in potential followed a definite cycle, even in normal people: 'In the fall and winter maximal positivity tends to occur around the new moon and maximal negativity around the full moon.'[13] In mental patients, this fluctuation was far more marked.

But from the point of view of this argument, perhaps the most significant observation was one made by Professor Y. Rocard of the Sorbonne in 1962. Rocard was studying dowsing – or water divining. Anyone who has ever dowsed will know that the forked twig or divining rod seems to dip of its own accord; the dowser feels nothing whatever. Rocard was sufficiently convinced by the claims of dowsers to test whether their power could be explained by earth magnetism. He discovered that underground water causes weak changes in the earth's magnetic field, which in turn produce an effect in the dowser's muscles, causing the rod to dip. Rocard tried many subjects who were not professional dowsers, and discovered that they could also detect the gradient in magnetic current. (My own experience is that nine out of ten people can dowse, once they have been shown how to use the rod, although some are far more sensitive than others.) In describing these experiments (in *The Cosmic Clocks*), Michel Gauquelin comments: 'Magnetic irregularities are not only caused by what is found underground; the sun and the moon also modulate the terrestrial magnetic field. Changes registered following solar storms and lunar transitions are of the same order of magnitude as those perceived by Rocard's subjects. . . .'[14]

Here, then, we have a key to the question of how our Stone Age ancestors came to know so much about the heavens. Primitive people are far better dowsers than civilized human beings; the aborigines of Australia can detect underground water without the need for a forked twig – it is often a matter of life or death for them. So it was for our remote ancestors. And if they could detect underground water, it follows that they could also detect the magnetic changes in the earth's field caused by sunspots and lunar activity – as well as by our closest neighbour, Venus.

I have attempted to present this argument in terms of scientific evidence, rather than, say, in terms of the kind of evidence Robert Graves presents in *The White Goddess*. But

---

* In this year, an alignment of all the planets on the same side of the sun should trigger heavy sunspot activity – and earthquakes.

Testing Candidates for the Position of Water Diviner on the
Metropolitan Waterboard. *A cartoon by Heath Robinson*

this is not because I would regard scientific evidence as more reliable than Graves's intuitions and mythological insights. Rocard, for example, concluded that dowsers could not tell the difference between still water and running water, and that they were not even sure whether the rod was dipping for water, for metallic objects, for iron ore, or even for rocks struck by lightning. No doubt this was true for the subjects tested by Rocard; but I have known a number of skilled dowsers who had no difficulty telling the difference between a metal pipe and an underground stream – and who could even distinguish between coins or other flat objects placed under a carpet. Weirder still – so weird that I almost blush to mention it in a book like this – is the fact that many dowsers can locate water from a map. In *Mysteries* I cite a case in which a Swiss dowser was able not only to locate a drowned body by means of a pendulum suspended over a map, but was able to follow its course down-river when the frogmen who went to investigate released it from the mud on the river bottom. In short, good dowsers are quite convinced that it is not simply a matter of response to the earth's magnetic field. The most general view seems to be that it is some form of mental 'radar' which can somehow distinguish between one substance and another.

How such 'radar' could operate on a map is obviously impossible to say. But our observations on the 'split brain' seem to offer at least a generalized clue. The experiment with the red and green lights – in which the right brain 'nudged' the subject by making him jump – reveals that the right brain has control over our muscular system – or, at least, over the so-called unstriped muscles that are not under the direct control of the will (which lives in the left brain). Dowsing *is* undoubtedly an involuntary muscular contraction. So it seems a reasonable guess that it is the right brain that picks up the change in the earth's magnetic field, and that responds to it by activating the muscles. In some dowsers, this response can be extremely powerful, sending them into convulsions – which suggests that there is more involved than a response to a minute magnetic fluctuation. Either the magnetism releases a violent muscular response – like an epileptic seizure – or the magnetic force involved is transmitted direct to the muscles.

At all events, it seems fairly clear that dowsing is the province of that 'other' person in the head, not the rational ego. And we know that the 'other person's' mental processes operate in a quite different way from those of the left brain, which proceeds logically, step by step, like a computer. The left brain is our instrument of survival; it operates by imposing a rational structure on the world, which has the effect of 'familiarizing' the environment. In order to navigate our way through the complexities of life, we need to know 'where everything belongs', so it imposes a kind of gigantic grid on the world, so everything can be neatly pinned down to definite squares. The very thought of things that cannot be pinned down in this way makes it nervous. For example, the idea of the infinity of space makes it dizzy, while it doesn't seem to bother the right brain – our 'intuitive self' – in the least.

But how does a baby recognize its mother's face? Not by some 'grid-like' process of analysis, but by grasping it as a whole. (Chomsky suggests that we learn language in

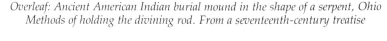

Overleaf: Ancient American Indian burial mound in the shape of a serpent, Ohio
Methods of holding the divining rod. From a seventeenth-century treatise

the same non-logical way.) The left brain cannot even begin to understand such a process. And neither, it seems clear, can the left brain understand what goes on in dowsing. Rocard's attempts to pin it down by studying earth magnetism are very creditable; but his rational net is bound to let all the smaller fish escape. Dowsing *seems* to involve a process of instantaneous intuition, like a baby's recognition of a face; and if this is accepted, then its results can no longer be regarded as contrary to common sense: only to the computer-like processes of the left cerebral hemisphere.

Let me, then, make clear what I am suggesting: that there is nothing contradictory in the knowledge of astronomy that seems to be displayed in monuments like Stonehenge and Carnac, even though these were constructed in the 'pre-scientific era'. One investigator of ancient monuments, Tim O'Brien, has concluded from his study of the Wandlebury earthworks near Cambridge that it shows the same astronomical knowledge as the outer ring of Stonehenge (with which he believes it to be contemporary); he finds evidence, not only that the builders knew the earth was round, but that they knew its circumference within 1 per cent. He theorizes that these builders were not the native inhabitants, but 'wise men' whom the Irish call Tuatha De Danaan, the children of Danu (or Anu). Unlike Erich von Däniken, O'Brien is not willing to believe that these beings came from outer space; he believes they were Sumerians (who worshipped a god called Anu), who journeyed from Sumer (southern Mesopotamia) via Scandinavia, Scotland and Ireland. I find his reasoning sensible and almost convincing – except that it leaves the basic question unanswered: where did Neolithic 'wise men' gain such a precise knowledge of astronomy? The 'dowsing hypothesis' answers that question. Ancient man *felt* the forces of the sun, the moon and the planets; he experienced them directly as an influence on the earth. There must have been then, as now, certain men who had an unusual level of sensitivity, so that they were deeply disturbed by magnetic changes caused by the heavenly bodies. Such men probably became priests, or were at least attached to the temple. (It is conceivable that the notion of the 'holy fool' originated in these 'lunatics'.)

Let me clarify my meaning with an anecdote. The travel writer Negley Farson, a friend whom I saw frequently in the late 1950s, lived in a house in north Devon overlooking the Bristol Channel. One day, as he was looking out of his bedroom window, he saw an aeroplane crash into the sea and vanish. While the sea was still foaming and bubbling, Negley quickly stuck a small piece of stamp paper on the window, so that it blotted out the bubbles. He then moved to another spot in the bedroom, and looked at the site from a different angle; again, he stuck stamp paper on the window. Then he went to the telephone and reported it to the police. With the aid of his two pieces of stamp paper – and the knowledge of where he had been standing when he stuck them on – the authorities were able to pinpoint the spot where the plane went down, and recover the wreck.

It was necessary for Negley to see – to have direct sensual experience – of the spot where the plane went down before he could fix his 'markers'. And if we suppose that the prehistoric priests also had this direct sensual experience of the sun and moon, and some of the planets, we can also imagine them using stones as markers. If Alexander Marshack is correct about the markings on the Dordogne reindeer antler, Cro-Magnon man wanted to memorize the exact positions of the moon over a certain period, and tried to represent these by arranging the marks in a sinuous line. The next obvious step was to do what Negley Farson did, and take 'sightings' on definite places on the horizon where the planets, sun and moon rose and set, and mark them with stones.

27

This is what Hawkins and Hoyle have suggested was done by the builders of Stonehenge. (Two Russian scientists, Vladimir Avinsky and Vladimir Tereshin, have recently gone even further, and announced their discovery that the arrangement of the stones contains basic information on the size of the planets, and that the estimate of the size of the earth and moon has a margin of error of only 1 per cent.)[15]

But why do we need to hypothesize that the priests had some intuitive sense of the movements of the sun and moon, since they could obviously lay out a calculator like Stonehenge merely by using their eyes? Because we need to explain why they wanted to do this in the first place. It is a different matter when we come to the age of scientific astronomy – say, after 600 BC – because by then a complex, if basically nonsensical, system of astrology had been developed, and men believed that the position of the planets foretold wars, plagues and the deaths of kings. But the ancient worship of the sun and moon was a matter of religion rather than of fortune-telling; and religion is based on feeling and intuition, not on theories.

We also need to account for the curious fact that sacred sites seem to retain their character of holiness over centuries, even millennia; men go on choosing them for temples and churches. An immense number of Christian churches – usually dedicated to St Michael and the Virgin Mary – are built on sites of Iron Age hill forts, which in turn were the sites of earlier temples dedicated to the sun god or mother goddess. However, if we accept.the premise that Stone Age 'dowsers' were sensitive to the forces of the earth – and their response to the sun, moon and planets – then the answer that suggests itself is that there is some peculiarity in the earth's magnetic field in such places. In an article on the source of the earth's magnetic field, two geophysicists comment that its direction 'varies sporadically from region to region, so that the field is seen to consist of irregular eddies'.[16] The word 'eddies' is of particular interest here, since modern dowsers like Guy Underwood insist that sacred sites like Stonehenge are distinguished by a field of force in the shape of a whirlpool. The Chinese referred to the earth force as the 'serpent force' – suggesting a coiled snake – and images of spirals are common on standing stones and ancient monuments from all over the world. (Glastonbury Tor, a sacred site of immense antiquity, even has a spiral ramp encircling the hill.) The source of the earth's magnetism is still unknown, although the most widely held theory is that the movements of molten masses of iron generate electric currents. G. O. Roberts, a Cambridge geophysicist, points out that fluid motion can generate magnetism if the motions possess a net 'helicity'. This is the way in which flow streamlines are twisted into right or left handed coils; net helicity is an imbalance, in which there are more coils of one kind than another. Again, the notion of coils is connected with earth magnetism. . . .

All this still leaves one major question unanswered: what did the monument builders *do* at their sacred sites? What was the final purpose of all this astronomical observation? The question, admittedly, belongs to the realms of anthropology rather than astronomy, and could therefore be regarded as irrelevant; still, for the sake of completeness, it seems desirable to make at least some suggestions.

There appears to be no foundation to the widespread notion that sites like Stonehenge were used for human sacrifice. It is true that the skeleton of a baby with a split skull was found at Woodhenge (two miles from Stonehenge), and this *could* have

'Old Sarum' (an ancient British stepped pyramid), Salisbury Cathedral and Clearbury Ring in alignment. The Old Sarum ley runs for $18\frac{1}{2}$ miles, passing through Stonehenge

Frankenburg
Camp

Clearbury
Ring

Salisbury
Cathedral

Old Sarum

Stonehenge

tumulus

N

been a sacrifice to dedicate the site. But this is very much the exception. The stories of human sacrifice originated with antiquaries like John Aubrey who surmised that the megalithic circles were built by the Druids – Celtic priests who were active from about 500 BC onward. The Druids *did* perform rites of human sacrifice – usually by burning – and no doubt Druids did use Stonehenge as a temple at this late date; but there is no evidence for human sacrifice there.

Professor Margaret Murray, author of *The God of the Witches*, died before Hawkins revived the astronomical theory of Stonehenge; but she would have found nothing in it to disagree with. Margaret Murray was convinced that modern witchcraft is a survival of an ancient cult of Diana – the Roman moon goddess – which dominated pre-Christian Europe. This religion identified the moon goddess with the earth goddess, and was basically a pagan fertility cult – that is, its main aim was to make sure that the earth became fertile every spring.

Where the builders of Stonehenge 1 are concerned, the only objection to this otherwise convincing theory is that they had no need to make the earth fertile, since they were not farmers. The evidence indicates that the worship of the moon goddess pre-dates farming – possibly by thousands of years. But certainly, there *is* evidence of an almost universal religion of the moon and the earth mother long before the coming of any of the present great world religions.

In the late 1960s, another theory of the purpose of the megaliths began to make headway among those who were interested in the mystery of the ancient monuments. In a book called *The View Over Atlantis*, John Michell suggested that not only were temples like Stonehenge built on 'eddies' of magnetic force, but that lines of smaller megaliths often trace the course of the 'magnetic currents' across the countryside. Michell called these lines of current 'leys', borrowing the word from a Hereford businessman, Alfred Watkins, who had coined it in a book called *The Old Straight Track* in 1925. But Watkins's leys were not lines of earth magnetism, only old, straight roads which he claimed to detect all over the English landscape. He thought that they were prehistoric trade routes, although he later speculated that they might have had some religious significance too. Michell amplified Watkins's suggestion into the theory that the religion of prehistoric man was basically concerned with this 'earth force', and that this is the true solution to the problem posed by the thousands of megaliths and stone circles which can be found all over the world. Respectable archeologists and geophysicists were quick to denounce Michell and his followers as members of the lunatic fringe. Yet, like the astronomical theory of Lockyer and Hawkins, his views are slowly beginning to edge their way into respectability.

Professor Euan MacKie explains in his preface to *Science and Society in Prehistoric Britain* that one of his reasons for writing the book 'is that there have been many works from the wilder fringes of science and scholarship on the general theme that one might describe as "the wonderful secret knowledge that our ancestors possessed and that we have lost"', and adds that 'such topics run the gamut from Stone Age computers through leys and mysterious power lines across the countryside to Atlantis and even to contact with visitors from other worlds'.[17] But MacKie's own book is also about the secret knowledge of our ancestors and Stone Age computers. And, as I have tried to show in previous pages, if we can accept that the megaliths were astronomical computers, then we are coming very close to the notion that our ancestors had powerful religious reasons for associating the earth, the sun and the moon. And if we accept that these people were basically savages, we are again faced with the mystery of why a primitive people needed sophisticated astronomical calculators. The 'dowsing hypothesis' is the only one that fits the facts. In which case, Michell's ley lines deserve to be classified with the Stone Age computer theory rather than banished to the 'wilder fringes' together with Atlantis and visitors from other worlds.

Michell's view, then, is that the Stone Age temples were places where men tried to enlist the aid of 'sacred forces' to ensure success in hunting or agriculture. We are now fairly certain that prehistoric cave drawings of animals were made as part of magical hunting ceremonies, the aim being either to direct the hunter to the herds of bison, or to lure the prey into the area of the hunters. If Michell is correct, then the ceremonies that took place in the megalithic circles had the same purpose. If, in fact, the 'altar stone' at Stonehenge *is* a sacrificial altar, then it was probably used for the sacrifice of animals rather than human beings, and the aim was to ensure the good will of the earth mother in protecting and feeding her children. Another aim may have been to ensure their health – many of the megaliths are traditionally used for healing.

If this view is correct, then astronomy began with something more than man's curiosity about the stars. It began with his sense of relationship with the earth, and his certainty that this relationship was influenced by the heavens. It began, in short, as a form of 'lunar knowledge' – of the kind of intuition that provided Edgar Allan Poe with the uncannily accurate insights of *Eureka*.

Art as sympathetic magic. Cave paintings of bison from the Dordogne, Southern France

CHAPTER TWO

THE PYRAMID MYSTERY

It is necessary to begin this chapter with some speculation that, at first sight, may seem to have little to do with astronomy.

In 1929, in the Topkapi Palace of Istanbul, a parchment map was discovered; it was signed by Piri Ibn Haji Memmed, an admiral (*re'is*) of the Turkish navy, who, like the English naval hero Drake, was basically a pirate. (In fact, he was beheaded in 1554.) The Piri Re'is map excited considerable attention because it included America; moreover, it showed Africa and America in their correct longitudes, a remarkable feat for the time, since navigators had no way of reliably determining longitude. (Latitude was easy enough, as it can be worked out by the position of the stars, which remain fixed in a plane as the earth turns on its axis.)

Stranger still, it showed Antarctica, which was not discovered until 1818. And when the map was examined by an American scholar, Captain Arlington Mallery, an even odder fact emerged. It showed a section of the coast of Antarctica – called Queen Maud Land – in the days *before* the bays were covered with ice.

From Piri Re'is's own notes on the map, it seemed that he based it on at least twenty earlier maps, including one of Columbus's (which has since vanished). But if some of these earlier maps showed the coast of Queen Maud Land free of ice, the consequences were, to say the least, startling. For radioactive dating of the sediment from the sea bottom in this area reveals that it was frozen solid from about six thousand years ago. So the original maps must have been made sometime before 4000 BC. Which was preposterous. Historians generally accept that seafaring over large distances began with the Bronze Age – about 2000 BC. We know the Sumerians sailed the Persian Gulf as early as 4000 BC, but compared to the Antarctic, this is a sheltered pond.

In the late 1940s an American professor of the history of science,

Charles H. Hapgood, decided to set his students to work on the Piri Re'is map. One of them pointed out that it resembled the navigation charts of the Middle Ages, known as portolans (meaning from port to port) and generally assumed to be rather inaccurate. Closer examination of these portolans revealed that they were far from inaccurate; it was simply that they had been drawn up on a different system from our modern one based on latitude and longitude. One of these – found in the Library of Congress and dated 1531 – showed Antarctica with considerable accuracy, and again showed the same lack of ice in coastal regions.

It became clear that most of these early maps had been based on a great many others – that is, they were compilations of many smaller maps of specific regions, all stuck together, so to speak. Inevitably, inaccuracies crept in; but the inaccuracies could be more revealing than the correct measurements. Hapgood's team discovered, for example, that the Piri Re'is map must have been based on maps drawn up by scholars of Alexandria, who at some point had incorporated the latest piece of scientific knowledge – the measurement of the size of the earth by the Greek astronomer and geographer Eratosthenes in the third century BC. Not surprisingly, Eratosthenes had made a small error – of about 5 per cent. But internal evidence showed that the Alexandrian scholars must also have possessed older maps that had been drawn without this 5 per cent error. In other words, Eratosthenes only re-discovered the size of the earth; it must have been discovered – far more accurately – so much earlier that the Alexandrians were unaware that the earlier maps were, in fact, more precise.

A Turkish map drawn up in 1559 not only shows a spherical projection that looks modern in its sophisticated accuracy; it also shows a great land bridge across what is now the Bering Strait – the strait between Siberia and Alaska. This land bridge *did* exist in the remote past, as we know from the similarity of animal remains in the two continents. But again, it was certainly more than six thousand years ago.

Perhaps the most startling discovery made by Hapgood's team was that a Chinese map, reproduced in Joseph Needham's great *Science and Civilisation in China* and dating from AD 1137, showed the unmistakable fingerprint of the same mapping technique as the Piri Re'is map.

All this evidence led Hapgood, in *Maps of the Ancient Sea Kings*, to advance a staggering hypothesis: in some distant epoch, possibly as long ago as ten thousand years (when our ancestors were supposed to be living in caves or primitive settlements), there was a highly advanced civilization that reached over the whole globe – from China to Antarctica, from Alaska to South America. In science and mapmaking it was more advanced than the later civilizations of Greece and Rome. It disappeared – perhaps suddenly, perhaps gradually – at such a remote period that it was totally forgotten. Hapgood has no doubt that a whole civilization could vanish like this. The more advanced the culture, he says, the more easily it could vanish without trace: how much evidence of modern New York would exist if it was destroyed by a hydrogen bomb? If this ancient civilization was so advanced, then its powers of destruction must also have been enormous. . . .

It must be emphasized that Hapgood is a serious scholar, not a popularizer or casual speculator. When Jacques Bergier – one of the authors of the famous *Dawn of Magic* – summarizes Hapgood's theories (in *Mysteries of the Earth*), he states that the land bridge across the Bering Strait must have existed about thirty thousand years ago, and that the original maps must date back at least fifteen thousand years. This is an absurd exaggeration. Neither is there any evidence in Hapgood's book that the maps must

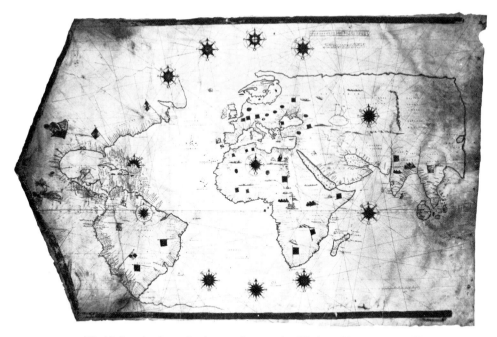

World chart in the style of a portolan map by Girolamo Verrazzano, 1540

have been made by means of flying machines – perhaps by the 'space visitors' popularized in the books of Erich von Däniken. Everything *could* have been achieved by the kind of science developed by the Greeks around 500 BC, and by a good navy. (Moreover, if the maps were originally drawn by 'space men', then there would also have survived – in Egypt, for example – the knowledge that the earth is not the centre of the universe, or even of the solar system; and all ancient astronomers, as we shall see, took the contrary for granted.)

Hapgood was unfortunate in that, in the year following publication of his book, von Däniken's *Chariots of the Gods* appeared, with its theory of 'ancient astronauts' and its mass of inaccuracy and misinformation. Von Däniken soon came under heavy attack from scholars of all kinds, and Hapgood's work became involved in a kind of guilt by association. Yet in a book devoted to criticism of von Däniken by experts in various fields,[1] A. D. Crown (a senior lecturer in Semitic studies) – after attacking von Däniken's views on the Piri Re'is map – remarks casually that he doubtless borrowed one of his illustrations from Hapgood's *Maps of the Ancient Sea Kings*: the implication being that Hapgood is reliable and respectable while von Däniken is not. This is no doubt true; but Crown is attacking the whole notion that the ancient maps pre-date Columbus – a position which Hapgood substantiates with a wealth of detail and calculation. In a recent revised version of his book published in England (1979), Hapgood remains unrepentant; and it must be admitted that, after more than a decade, his evidence stands as solid as ever.

The relevance of all this to our argument should now be clear. Seafaring is impossible without a knowledge of the stars. So if Hapgood is correct, then a detailed knowledge of astronomy must pre-date *all known civilizations*.

And how would this ancient civilization have determined the size of the earth? Presumably by using the same method as Eratosthenes. The Greek scientist, who was based in Alexandria, heard about a deep well in the town of Syene (now Aswan), where the sun was reflected at midday every midsummer day. (Their midsummer was 21 June – not, like ours, 24 June.) Syene is, of course, in the tropics, and the tropics are the only area of the earth where the sun can stand directly in the centre of the heavens; in all other latitudes it will be seen to the north or south.

In Syene, then, a tower would not cast a shadow at midday on 21 June. But at Alexandria, it did. Eratosthenes only had to measure this shadow, and determine the angle of the sun's rays to the tower. It proved to be $7\frac{1}{2}°$. Simple mathematics told him that if the earth is a globe, then the distance from Syene to Alexandria must be $7\frac{1}{2}°$ of the earth's circumference. And since he knew that distance – 5000 stadia (or 500 miles) – he was able to calculate that the earth must be 24 000 miles around its equator. Since the modern measurement is 24 860, there is an error of less than 5 per cent.

Now oddly enough, Hapgood found that Syene played an important part in the construction of the Piri Re'is map. This map was only a fragment of a much larger one; it showed the west coast of Africa and east coast of America. A mapmaker needs some kind of 'grid' to help him – a modern mapmaker obviously uses the lines of latitude and longitude. The ancient mapmakers had evidently chosen a different method, presumably because they realized that a square grid would produce distortion (with the north pole spread along the top of the map and the south along the bottom). Hapgood, with the help of several mathematicians, worked out the method used by the old mapmakers. They had chosen a centre, drawn a circle round it, subdivided the circle into sixteen segments, then drawn various squares with their vertices touching the edge of the circle – a complicated method, but one which worked well enough. But the original centre of the circle was on the missing part of the map, off to the right. By extending various lines on the map back to their source, Hapgood deduced that the original centre was in Egypt, possibly in Alexandria. But more accurate calculation showed that Alexandria was too far north; the lines actually converged at Syene.

Syene would have been a logical centre for the Alexandrian mapmakers after Eratosthenes. But if, by any chance, the ancient mapmakers – before Eratosthenes – had also used it, then it would imply that they had used the same well to determine the size of the earth. . . .

There is, of course, no direct historical evidence for this. But there is an interesting piece of indirect evidence. The Egyptians' knowledge of astronomy was considerable, even as early as the third millennium BC; and they also knew the exact size of the earth. The Greek grammarian Agatharcharides of Cnidus was tutor to one of the Ptolemy kings of Egypt at the end of the second century BC. He was told that the base of the Great Pyramid of Cheops was precisely one-eighth of a minute of a degree – that is, the length of that part of the earth's circumference. A minute – in geometry – is one-sixtieth of a degree. A few moments with a pencil and paper will show that the length of the pyramid's base – just over 230 metres – multiplied first by eight, then by sixty, then by 360, does, in fact, give a distance of just under 40 000 kilometres or just under 25 000 miles, a remarkably precise estimate.

It is true, of course, that Agatharcharides lived a century after Eratosthenes, and so

The Piri Re'is map of 1513. The Drake Passage is omitted and the coast of Queen Maud Land (bottom right) is shown free of ice

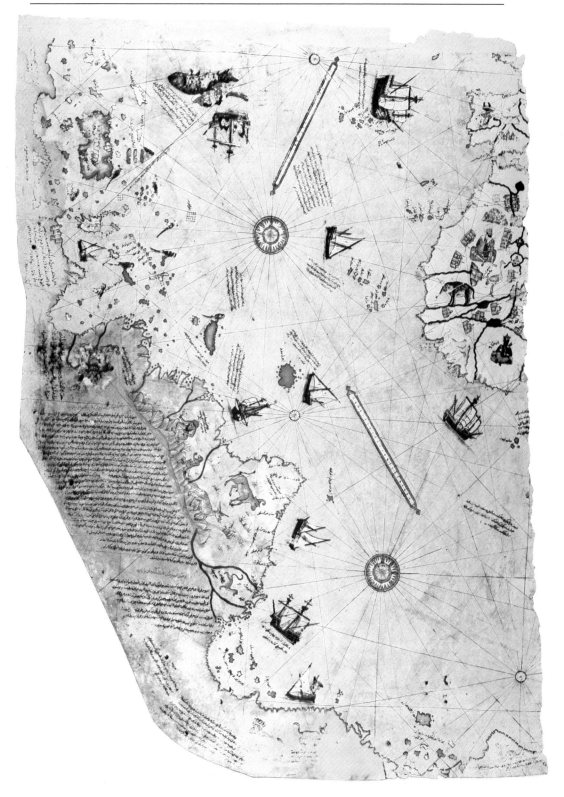

could easily have worked out the ratio of the base of the pyramid to the earth's circumference for himself. But he was reporting it as an ancient tradition about the pyramid, not as a modern measurement; just as the Greek historian Herodotus, two centuries earlier, was reporting ancient tradition when he said that the surface of each face of the pyramid is equal to the square of its height.

So the Egyptians knew the size of the earth long before Eratosthenes – their knowledge had been long forgotten by his time. (We shall look at more evidence for this in a moment.) But then, Syene and Alexandria *were* in Egypt. In fact, the length of Egypt – which was also precisely known in the third millennium – was measured between the Nile delta (on which Alexandria stands) and Syene, on the great cataract where the Aswan dam stands today. So there is good reason why mapmakers, long before the time of Eratosthenes, should have used Syene as a centre point; they had probably used the same well as he had to determine the size of the earth. Or, alternatively, they may have inherited the knowledge from the much older civilization which Hapgood postulates. . . .

When the Greek traveller Herodotus visited Egypt in around 440 BC, he went to see the Great Pyramid, which was already more than two thousand years old, and questioned the priests about it. He was told that it had been erected on the orders of a wicked king called Cheops, who had closed the temples and forced his subjects to labour at its construction. Herodotus does not actually mention *why* Cheops had the pyramid built, but he adds that the vaults under the pyramid were intended for the king's own use – that is, as a tomb; he adds that they were constructed on an underground island with water from the Nile flowing round it.

In the seventh century AD, Alexandria was captured by the Arabs. Two hundred years later, Abdullah Al-Mamun, the son of the great Harun Al-Rashid, the Caliph of the *Arabian Nights*, commissioned scholars to construct a complete map of the earth, and another of the heavens. (It would be interesting to know whether this map showed America and the Antarctic; unfortunately, it has disappeared.) Hearing a legend that the Great Pyramid contained a secret chamber that was full of ancient maps, Al-Mamun determined to find it. In 24 BC, the Roman geographer Strabo had reported that there was a hinged door in the north face of the pyramid; but its site had long been forgotten. Al-Mamun decided that he would tunnel his way straight through the masonry. The stones were heated by huge fires, then water was thrown on them to make them crack, so they could be broken apart by battering rams. His men hacked a tunnel for a hundred feet into solid masonry. Luck was with them; they heard a muffled crash to the left, as something fell. They changed the direction of the tunnel, and broke through into a narrow passage, less than four feet high. Scrambling up this passage, they discovered the missing 'secret entrance' of Strabo. In considerable excitement, they hurried back down the passage, hoping it led direct to the secret chamber. It descended below ground level, into the bedrock. And there, to their disappointment, they found only a tiny room full of debris. On its far side, a horizontal passage led to a dead end. A well shaft had been dug in its floor.

The Arabs now discovered what had caused the dull thud – a stone had fallen from the ceiling. It was shaped like a prism, and in the hole it left they could see what looked like a flat rectangular door. Torch marks in the passage revealed that it had been visited in previous epochs – probably by the Romans; but the earlier visitors had known nothing of this 'secret door' which seemed to lead upwards. The Arabs tried to chip at its edges; it blunted their chisels. But the surrounding rock was less hard; they hacked and chiselled their way through it for six feet, only to find another granite 'door' – or plug – behind the first. When they had burrowed past this, there was another. Then they encountered a limestone plug. Eventually, they found themselves scrambling up another narrow passage; it led to an equally low horizontal passage, and this in turn led them to a small room with a pointed ceiling; again, this was empty – as was a niche in the wall which looked as if it might once have contained a statue or upright sarcophagus. Because Arab women were sometimes buried in tombs with gabled ceilings, the room became known as the Queen's Chamber. Suspecting a tunnel behind the niche, the workmen removed its rear wall; but nothing was found. The Queen's Chamber was a dead end.

They retraced their steps along the horizontal passage, to the point where it began to slope. Holding up their torches, they saw another 'entrance' above their heads: not a vertical tunnel, but another sloping passage. This was, in effect, a continuation of the lower ascending passage. When they climbed into this passage, they found themselves in the most impressive part of the pyramid so far, a vast gallery. It stretched up far above them – the ceiling was twenty-eight feet overhead. Scrambling up the smooth surface, aided by small ramps placed on either side at regular intervals, they found a low door leading into an antechamber. Beyond that, a short passage led into another chamber – much larger than the Queen's Chamber. The workmanship of the polished granite blocks left no doubt that this was a room of major importance. It was thirty-four feet long, seventeen feet wide, and nineteen feet high. The ceiling was flat. But it contained nothing but a huge lidless chest – or sarcophagus – of polished brown granite. This was empty. This room became known as the King's Chamber. And, once again, it was a dead end.

The secret of the Great Pyramid, it seemed, was that it contained nothing – no body, no treasure. Had it been looted? That seemed impossible, for the granite plugs that sealed the ascending passage had obviously never been disturbed.

The sun's rays dictating the shape of the pyramids. A late Victorian theory

For another eight hundred years after Al-Mamun, the pyramid remained undisturbed – apparently having yielded up its disappointing secret. Arabs of later generations stripped off the shining outer casing of limestone and used it for building, leaving the rough stones exposed. Then, in 1638, an Oxford mathematician named John Greaves set himself the task of calculating the basic measurement the Egyptians had used to construct the pyramid. And at the junction of the sloping ascending passage and the horizontal passage leading to the Queen's Chamber, he noticed a lop-sided stone, covering a narrow shaft. This proved to be a kind of well and was partly filled with rubble; the air was so bad that Greaves was forced to abandon his attempt to find out where it led. Two centuries later, an Italian named Caviglia discovered that this 'well' led back into the descending passage.

Greaves's discovery only deepened the mystery of this absurd – and apparently purposeless – construction. He was particularly puzzled by the twenty-eight-foot high 'Grand Gallery' that ascended to the King's Chamber; it was too steep to be a hall, and too slippery to be a ramp or stairway. And why was it 'hidden' at the junction of two low, narrow passageways? Greaves discussed this, and other mysteries of the pyramid, in a book called *Pyramidographia*. He reached no conclusion, but at least his measurements enabled Sir Isaac Newton to work out the basic unit of measurement used by the Egyptians. He concluded that there had been two: a 'sacred cubit' of about 25 inches, and an ordinary cubit of about 21 inches. Sir William Harvey, discoverer of the circulation of the blood, also pointed out that the King's Chamber must have some form of ventilation. (In the 1830s, this was confirmed when a man climbing the pyramid discovered two vents which led down to the King's Chamber; when cleared, they enabled the chamber to maintain an absolutely unvarying temperature of 68°F.)

In 1765, an American traveller called Nathaniel Davison discovered a tiny 'attic' above the King's Chamber – a mere three feet high; and in 1837, an English colonel, Richard William Howard Vyse, found three more 'attics' above this – one of which contained a hieroglyphic inscription containing the name of the Pharaoh Khufu – or Cheops. (By this time, the famous Rosetta Stone, containing inscriptions in Egyptian and Greek, had been deciphered by Jean François Champollion, enabling scholars to read hieroglyphs.) Howard Vyse concluded that the purpose of these 'attics' was simply to reduce the immense pressure of masonry on the ceiling of the King's Chamber in case of earthquakes – one of which had already cracked some of the blocks.

In 1798, Napoleon landed in Egypt with an army of 25 000 men, and a small army of French scholars. After defeating the Mamelukes at the Battle of the Pyramids in July 1798, Napoleon gave his scholars permission to try to discover the secret of the Great Pyramid. Their results were disappointing. The place was full of bats – their numbers had increased in the past century – which swooped about and attacked them as they scrambled up and down the narrow passages, half stifled by heat and dust. But at least their measurements were the most accurate so far. By clearing the debris around the base, they were able to establish that each side was just under 231 metres long. And a young scholar named Edmé François Jomard clambered to the top of the pyramid – it took him a full hour – then carefully measured each step on the way down, obtaining a height of 144 metres (or 481 feet). The length of the slanting side of the pyramid (known as the apothem) proved to be just under 185 metres. The original pyramid, of

Entering the Grand Gallery in the Great Pyramid. By Luigi Mayer, 1802

course, was covered with the outer limestone casing, so figures for its original height and length could only be approximate. (The later discovery of two of these limestone blocks, buried under rubble, enabled their thickness to be taken into account.)

But was all this of any real significance? In other words, what did it *matter* how big the Egyptians of 2500 BC had made the pyramid? Parts of the inside were finished very roughly indeed – suggesting that the original designer had died while it was under

Cross section demonstrating the use of both passages to observe the pole star

construction, and that the rest had been botched. Did not this suggest that its actual size – or orientation – was no clue to its 'secret'?

Jomard and his colleagues, however, made a discovery that contradicted this view. The pyramid is ten miles away from Cairo, which is at the base of the Nile delta – a triangle of streams running north into the Mediterranean. Its four sides point to the four points of the compass. Jomard and his colleagues established that this alignment – north, south, east and west – was extremely accurate. Engineers who aimed at this kind of accuracy knew precisely what they were doing. If diagonals were drawn from the pyramid – lines to the north-east and north-west – they neatly enclosed the whole Nile delta. The meridian line drawn from the north of the pyramid sliced the delta into two exact halves. All of which suggested that the Egyptians were expert surveyors, and had placed the pyramid with extreme care. Since the distance from the Mediterranean to the pyramid is 158 kilometres, it was quite clear that the Egyptians had some accurate method of measuring long distances – apart from having a man trotting over the ground with a piece of string.

Jomard found in Strabo and Diodorus Siculus the information that the apothem of the Great Pyramid was precisely one stadium long. The Greek stadium had been 185.5 metres long, which was almost the length *he* had measured for the side. The classical authors also asserted that a stadium was the six-hundredth part of a geographical degree of longitude – they had worked this out, of course, after Eratosthenes had discovered the circumference of the earth. Jomard quickly worked out that the apothem of the pyramid, multiplied by six hundred, *was* indeed the length of a degree of longitude in Egypt.* Was it conceivable that the Egyptians *knew* this, more than two thousand years before Eratosthenes?

Jomard also followed up the report of Agatharcharides that the length of each side of the base was an eighth of a minute of a degree of longitude; again, it worked out exactly. Now the French metre *is* a distance worked out with a knowledge of the

Napoleon visiting the pyramids of Giza during his Egyptian campaign, 1799

earth's size – it is supposed to be precisely one ten millionth of the distance from equator to pole. And if Jomard was correct, then it looked as if the Egyptians had used a measure also based on the earth's size – in this case, 1 divided by 216000.

Jomard's colleagues had no patience with these weird speculations. They pointed out wearily that the Egyptians had no knowledge of geometry – it was invented by the Greeks – and that they had hardly more knowledge of astronomy. Even a carving of an ancient zodiac, found in a temple at Dendera, failed to convince them. If the ancient Egyptians were interested in the sky, they said, it was solely for astrological reasons – because they thought the stars foretold the future. The kind of knowledge Jomard was talking about would have to wait for another two thousand years.

Jomard's views had been forgotten by the time another mathematician, John Taylor, began to study the problems of the pyramid, stimulated by the explorations of Howard Vyse in the 1830s. Taylor made a model of the pyramid and treated it as a mathematical puzzle, trying out various possible solutions. Did the pyramid really embody strange mathematical formulae, like the one Herodotus mentioned? Why, for

---

* The size of a degree varies over the earth's surface, since the earth is not an exact sphere – it bulges at the equator.

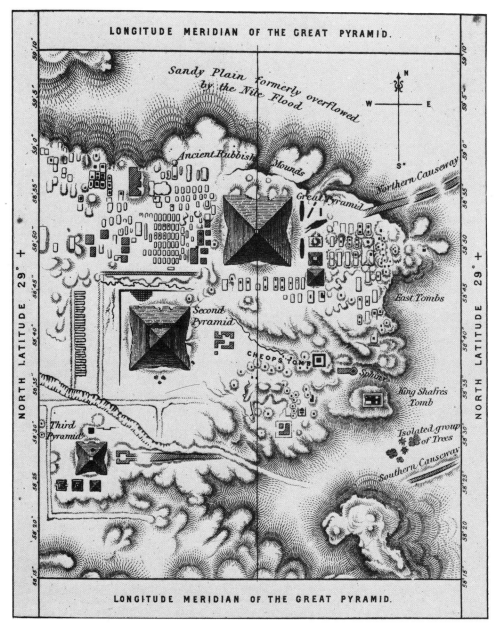

Aerial view of the complex of buildings at Giza. All the pyramids are oriented to the four cardinal points of the compass

example, had the builders chosen to slope the sides at 51° 51', instead of a more sensible round figure like 45° or 60°? The logical solution was that it *had* to be at that angle if it was to have a certain precise length of base and an equally precise height. Taylor brooded on this problem of the relation of base to height until he was struck by a revelation. Using the latest and most accurate measurements, he discovered that the length of the base, divided by twice the height, equals the quantity π – pi, the relation of the diameter of a circle to its circumference. This sounds unimportant enough until we realize that it is, so to speak, the radius of our earth compared to its circumference at the equator.

In other words, the pyramid contains the proportions of half a globe, the base representing the equator, and the height the distance from the north pole to the centre. And if the ancients were telling the truth when they reported that its height was one six-hundredth of a degree of longitude and its base one-eighth of a minute, then the Great Pyramid could be a representation of one half of our planet. But why give it flat sides? Obviously, because the Egyptians could not build a gigantic hemisphere or geodesic dome. Instead they embodied their results in a pyramid. . . .

Taylor was also excited by Newton's discovery that the sacred cubit is within a hair's breadth of 25 British inches, as well as by his own observation that one side of a British acre is equal to precisely one hundred of these cubits. It led him to speculate that the British inch could be a measure that had survived from the days of the ancient Egyptians. (It also led a sect called the British Israelites to advance the notion that the British are the ten lost tribes of Israel – but these strange speculations have no place in this survey.[2])

Unfortunately for the science of pyramidology, Taylor was also a man of strong religious beliefs, who accepted Archbishop James Ussher's view that the world was created in 4004 BC, and that the Flood occurred in 2400 BC. Since his own estimate of the age of the pyramid was that it was built in 2100 BC (he was four centuries out), it seemed impossible that men could have devised such a structure a mere 300 years after the Flood without divine aid. And if God had guided the architect of the pyramid, then he must certainly have intended it to incorporate the divine revelations of the Bible . . . . Taylor's work spawned a host of pyramid cranks who measured the various passages inch by inch, interpreted them as symbolic measures of time, and found major events of world history to have been foretold by the measurements.

Taylor's friend – and disciple – Charles Piazzi Smyth, Astronomer Royal of Scotland, was one of these religious interpreters, and his book on the pyramid explains with staggering simplicity why the Grand Gallery is so different from the narrow passage by which it is reached: its beginning symbolizes the birth of Christ. The Second Coming will occur around the year 1911. . . . But Smyth was also a conscientious investigator; he went to Egypt, and carefully re-measured the whole pyramid. His measurements confirmed that Taylor was correct in finding the figure π in the relation of height to base. This result was derided by fellow scientists and scholars; but an ancient Egyptian mathematical document, the Rhind Papyrus, seems to prove him right. It dates from 1650 BC – admittedly eight hundred years after the Great Pyramid – and contains a value for pi of 3.16. (It was not worked out to three figures – 3.142 – until the sixth century AD.)

In 1883, nineteen years after Piazzi Smyth's highly successful book on the Great Pyramid, an astronomer finally made the suggestion that now seems so obvious, but which at the time seemed absurdly far-fetched – that the pyramid was intended as an observatory. He was Richard Anthony Proctor, and he entitled his book *The Great Pyramid, Observatory, Tomb and Temple*. Proctor noted that the Greek philosopher Proclus reported that the pyramid had been used as an observatory *before* its completion. A number of Arab historians had also stated that it was designed as an observatory. No one had paid any attention to these comments for the obvious reason: how *could* a pyramid be used as an observatory – particularly if, like the Great Pyramid, it had no obvious windows or other entrances?

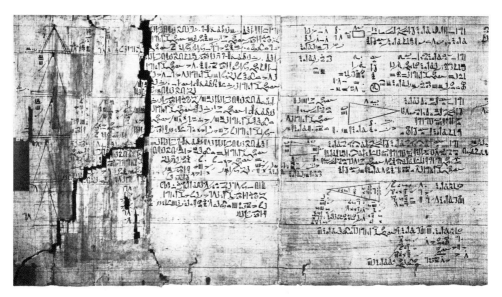

*The Rhind Ahmes mathematical papyrus, 1650 BC. This contains a value for
pi – which is the relation of the Great Pyramid's height to its base*

What Proctor realized about the pyramid was what Norman Lockyer later realized
about Stonehenge – that one of the prime necessities for an agricultural civilization is
an accurate calendar, particularly if their religious worship is also based on the gods of
heaven and earth. And an accurate calendar must involve precise observation of the
movements of the moon and stars. What they would have needed, to begin with, was
a long narrow slot, pointing due north (or south), through which the passage of the
stars and planets could be accurately observed. It could be compared to one of those
clocks where the time appears as numbers in a slot – except that the pyramid slot
would have to be vertical, not horizontal.

The first thing the engineers would have done, says Proctor, is to determine true
north, then align some kind of tube on it. Nowadays we would simply point a tube at
the pole star; but in those days, it was not in the same position. (This is because the
earth's axis has a wobble, owing to the gravity of the sun and moon; so a long rod stuck
through the centre of the earth, and emerging at either pole, would slowly describe a
small circle in the heavens. It causes the phenomenon we call precession of the
equinoxes.) But this presented no real difficulty. The stars all appear to describe a
semi-circle above our heads, from horizon to horizon. Those directly overhead
describe the largest circle; those near the pole, the smallest. The true north pole is at
the centre of the smallest circle. In the days when the pyramid was built, the star
closest to the pole was Alpha Draconis. If the Egyptians *had* aligned their 'telescope' on
this to find true north, the telescope would have pointed at an angle of 26° 17'. And,
oddly enough, this is precisely the angle of the 'descending passage' to the earth.

Understandably convinced that he was on the right track, Proctor went on to work
out how the pyramid could have been used as a star-clock. Once they had aligned their
building on true north by starting below the ground (as the descending passage does),
they would build their 'slot' pointing in the opposite direction. And what was this slot?
Obviously, the Grand Gallery, that baffling structure which now slopes upwards to

the King's Chamber. The ascending passage leading to the Gallery slopes upwards at the same angle of 26° 17'. Why should that be? Proctor's knowledge of modern observatories provided the solution. There must have been a reflector at the junction of the two passages, so that the light coming down the descending passage would be reflected at the same angle up the ascending one. A pool of water would serve that purpose ideally. And, significantly, the stones at this point in the descending passage are harder, smoother and more closely joined than elsewhere.

But a slot pointing at the sky would have one obvious disadvantage: it would only cover the small area of the sky that it points towards. To see stars below this narrow band, a man would need to walk up the 'telescope'. And this, says Proctor, is why there are slots along the ascending ramp – to hold seats, at various levels. And what about stars overhead? To see these, it would be necessary to remove stones from the roof of the 'telescope'. And again, the roof stones on the grand gallery – whose walls slope inwards – *are*, in fact, individually removable.

Of course, the structure would become useless as an observatory once the builders had got beyond the top of the Grand Gallery, which is now in the heart of the pyramid. But this would have taken ten years or so – ample time for the astronomer priests to make accurate star tables and maps. It would have served its purpose.

In 1890, seven years after the publication of Proctor's book (which was generally ignored), a young astronomer named Norman Lockyer took a holiday in Greece and, looking at the ruins of the Parthenon, wondered whether it had been aligned astronomically. He was thinking, he said later, of the fact that the east windows of many English churches face the exact place of sunrise on the day of their patron saint – for example, churches of St John the Baptist face north-east.

Whether Lockyer had heard of Proctor's work on the Great Pyramid is not clear from his book *The Dawn of Astronomy* (1894). But at some point, it struck him that the temples of Egypt had been so carefully measured and documented by archeologists since Napoleon's visit that they would afford him a wealth of basic material for an investigation of the astronomy of the ancients. He went to Egypt, examined the great temple of Amon-Ra at Karnak, and concluded that it was so constructed that the sunlight on the longest day of the year – the summer solstice – would penetrate direct to the inner sanctuary. He went on to explore and measure many other temples. The result was the remarkable *Dawn of Astronomy*, in which he shows in detail how the Egyptian temples were oriented according to the movements of the heavens. Oddly enough, he only makes passing references to the Great Pyramid, although he remarks that 'it is impossible to doubt that these structures [the pyramids] were erected by a people possessing much astronomical knowledge'.[3]

Lockyer made one observation that certainly makes it 'impossible to doubt' his statement. If the temples had long passageways that were aligned on stars – as the descending passage of the Great Pyramid was aligned on Alpha Draconis – the priests would observe over a couple of centuries that they were ceasing to serve their purpose as calendars, for the precession of the equinoxes – the wobble of the earth's axis – means that the stars slowly change their position in the sky. (They also rise twenty minutes later each year, for the same reason.) So a temple would have to be re-aligned periodically. And that is precisely what *has* happened again and again – the temple at Luxor has four such alterations.

After about 3200 BC, the 'dog star' Sirius became the most important star in the Egyptian heavens. This was because it now rose just before the sun (at dawn) at the beginning of the Egyptian new year – when the Nile began to rise. And since the Nile flood was the most important event in the lives of Egyptian farmers, the 'dog star' (so called because it is in the constellation Canis) became the god of the rising waters. The temple at Dendera – where the famous zodiac was found – was oriented to the dog star, so that on the first day of the heliacal (dawn) rising of Sirius, the light was channelled along a corridor, to fall on the altar of the inner sanctum. This was another observation made by Lockyer in *The Dawn of Astronomy*. (In the same book, he refers briefly to the solar alignment of Stonehenge – an idea he was to develop ten years later in *Stonehenge and Other British Stone Monuments* – a work that, like his speculations on the Egyptian temples, failed to arouse any great interest.)

This temple of Hathor at Dendera had been causing controversy since the time of Jomard. Jomard had claimed that its zodiac showed the constellations in different positions from later times, and that it therefore proved the antiquity of the astronomical tradition in Egypt. But an inscription found in the temple dated it to the first century BC. (And the use of the familiar Greek signs for the constellations – the Crab, the Scales, etc. – revealed that it was carved at least two thousand years later than the Great Pyramid.) In the twentieth century, the Alsatian philosopher R. A. Schwaller – who devoted much of his life to demonstrating the scientific and spiritual achievement of ancient Egypt – showed that the zodiac nevertheless contains internal evidence of the temple's antiquity. It consists of *two* roughly superimposed circles of constellations, one centred on the geographical north pole, the other on the true north pole, the pole of the ecliptic (the point the earth's axis *would* point to if it didn't wobble). The zodiac has its east-west axis going through Pisces, showing that it was constructed in the Age of Pisces, about 2100 years ago. (As everyone knows, we are now about to enter the Age of Aquarius.) But two hieroglyphs on the edge of the zodiac suggest another axis that passes through the beginning of the Age of Taurus, more than four thousand years earlier. This indicates that the Egyptians knew about the precession of the equinoxes, and that the religious tradition enshrined in the temple at Dendera dates back a further four thousand years.* Certainly, archeological evidence reveals that the present temple stands on the site of an earlier one.

In recent years, Professors Livio Stecchini and Georgio de Santillana have argued strongly in favour of ancient Egyptian knowledge of precession of the equinoxes. Stecchini has written a long appendix to Peter Tompkins's *Secrets of the Great Pyramid* in which he summarizes all the evidence that the builders of the pyramid – and of the temples of Karnak and Luxor – not only knew that the earth was round, but also knew its precise dimensions, and the fact that it is flattened at the poles; they could also measure the precise length of the year, and had mastered the system of map projection to reduce a spherical surface to flatness. If Hapgood is right, then it is conceivable that they inherited this knowledge from the far older civilization which preceded them. All of which, it must be confessed, adds a certain plausibility to the theory advanced by a popular writer on ancient mysteries, André Tomas, that the Egyptians derived their knowledge of astronomy from the survivors of Atlantis.[4]

What does this bewildering mass of evidence indicate? One thing seems clear: that if the Great Pyramid was built as a kind of super-observatory – and the evidence is very

* It is only fair to note that John Anthony West, whose *Serpent in the Sky*, Wildwood House (1978), is the fullest account of Schwaller's philosophy so far, remains unconvinced by this particular argument, arguing that the two superimposed circles are too irregular to prove anything of the sort.

The zodiac at Dendera. The overlapping constellations mark different dates in the past and indicate that the ancient Egyptians knew about the precession of the equinoxes

convincing indeed – then the priests of the reign of Cheops must have considered the undertaking necessary in order to advance their scientific knowledge. Like the first moon rocket, it was an immensely costly venture, which was felt to be worthwhile for the knowledge it would uncover. Once that knowledge had been gained, the pyramid could be completed and sealed. This notion is supported by the mystery involved in the granite plugs. If they were slid into place – in the ascending passage – after the pyramid had been completed, then someone must have been trapped inside the pyramid to do it, and a skeleton – or skeletons – would have been found. It has been argued that the man could have escaped by the well – in which case, how was it half-filled with rubble? But if the plugs were slid into place *after* the priests had finally achieved their purpose – a detailed sky map, with precise times of the rising and setting of stars and planets – and *then* the pyramid was completed, the problem would not arise. The pyramid *had* to be completed, because its precise dimensions also encoded their observations of the size of the earth, as Piazzi Smyth demonstrated. *If the pyramid was used as a pharaoh's tomb, then his sarcophagus must be in a secret chamber underground, and is still waiting to be discovered.*

If the theory sounds far-fetched, this is mainly because we are unable to imagine ourselves into the world of the ancient Egyptians. They were a deeply religious people, and their lives were dominated by the seasons and by the earth. Their temples, like Stonehenge, were 'star clocks'. For them the mystery of the universe

Egyptian priests observing the stars from the Grand Gallery. From La Nature, *1891*

stretched above them in the heavens. To understand the mind of God – or the gods – they had to understand the clockwork of the sky. At some point, they discovered the precession of the equinoxes, which must have struck them as the strangest mystery of all. Why had the gods introduced this slight irregularity into the heavenly clockwork? If their knowledge of astronomy had included the realization that the sun, not the earth, is the centre of the solar system, they might have put two and two together – the flattening of the earth at the poles and the 'wobble' – but they had no such knowledge. So the 'wobble' was a mystery, a hint of what went on in the mind of God.

Herodotus alludes to the religious mysteries of Egypt, and says that he will keep silent on such sacred matters. This was not because he venerated the religion of the Egyptians – as a Greek he had his own gods – but because at the heart of contemporary

Greek religion, too, were its sacred mysteries. The mysteries of Greece were associated with deities of the earth – like Demeter, the harvest goddess, and Dionysus, the god of wine; the greater mysteries were celebrated with great secrecy, and to be an initiate was a high privilege.[5] More than that: it guaranteed salvation – an eternity spent in the Elysian fields. The Egyptian mysteries were undoubtedly as important and as sacred; they were intended to impart some ultimate spiritual knowledge.

Speaking of the Greek mysteries, Erich Kahler says: 'There is another current of Greek religion, a darker and more emotional one that had great appeal for the people. . . .This religion consists in the cults of chthonic deities, that is to say, deities of the depths of the earth and the sea and of the powers of vegetation and regeneration. . . .'[6]

In short, we are again speaking about the strange, primitive religion of the earth

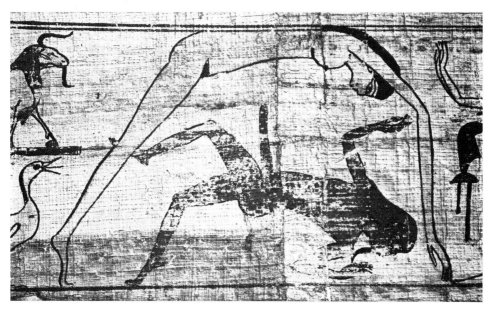

Nut, goddess of the sky, and Geb, god of the earth. The creation myth of ancient Egypt reflects the Great Religion of Neolithic times

mother and her peculiar powers – the powers that could turn men into diviners and magicians. 'The service of the mysteries began with an initiation and culminated in ecstasy, in rapture and madness, in a "sacred marriage" with the deity from which the human being was supposed to emerge reborn. . .', says Kahler.

This is the part of the problem that has been overlooked by most writers on Egyptian religion – including Tompkins, in his chapter 'Temple of Initiation'. The 'mysteries' were concerned not merely with the sky gods, but with the dark gods of the earth. The sky was important because it contained powers that could influence the earth – and that influence could be *directly sensed* by people who could tune in to it. If the Great Pyramid was used as a temple of initiation, then these were the mysteries that were celebrated there – a version of the same mysteries that were celebrated at Stonehenge, Carnac, Silbury Hill and Callanish.

It is difficult to describe the essence of these mysteries, but we can catch a glimpse of it in a passage from a book by Knud Rasmussen – the great Arctic explorer – on the Iglulik Eskimoes; an Eskimo shaman, or inspired priest, described how he became a shaman:

I soon became melancholy. I would sometimes fall to weeping and feel unhappy without knowing why. Then for no reason at all would suddenly be changed, and I felt a great inexplicable joy, a joy so powerful that I could not restrain it, but had to break into song, a mighty song with room for only one word: joy, joy! . . . And then in the midst of such a fit of mysterious and overwhelming delight I became a shaman, not knowing myself how it came about. But I was a shaman. I could see and hear in a totally different way. I had gained my enlightenment, the shaman's light of brain and body, and this in such a manner that it was not only I who could see through the darkness of life, but the same bright light also shone out from me, imperceptible to human beings but visible to all spirits of earth and sky and sea, and these now came to me to become my helping spirits.[7]

This ability to *see* the invisible, to 'hold converse with spirits', is a universal feature of all shamanistic religions, from Japan to Iceland. And the 'spirits of earth and sky and sea' are the same chthonic spirits that Kahler talks about in discussing the Greek mysteries.

Another key to this riddle is supplied by the archeologist Gertrude Levy in her book *The Gate of Horn* discussing religious conceptions of the Stone Age. The important point she makes in this book is that what might be called 'the Great Religion' begins a very long time ago – 'This story begins with the retreat of the icefields after the last and greatest period of cold ever known to these latitudes'[8] – and that it was a religion that was at once a cult of the dead and a cult of resurrection. This is a matter that I have hardly touched upon so far; yet it is of crucial importance to the argument. The predecessor of modern man – Cro-Magnon – was Neanderthal man, an ape-like creature who was killed off by our ancestors at least 25 000 years ago. Yet this 'missing link', who must have struck Cro-Magnon man as little more than a gorilla, carefully buried his dead with weapons, ornaments and offerings of food. One Neanderthal skeleton found at La Ferrassie showed signs of bandaging of the skull – an odd anticipation of the Egyptian custom. It seems, then, that the Neanderthals had some notion of a life hereafter. They undoubtedly had magical rites connected with hunting – animal bones and skulls smeared with red ochre have been found in ritual sites in the depths of caves. And since, among primitive people, religion and magic are inseparable, we can assume that their religion was more than a mere cult of the dead.

Gertrude Levy traces this dual religion – of the earth mother and death – down through Neolithic times, the age of the great megaliths, through Egypt, Sumer, Palestine, even Central America, and ends with a section on Greece and the mystery cults. And, at this point, she sees the Great Religion (my expression, not hers) fading as the mysteries develop into Greek drama, and into the philosophy of Heraclitus, Empedocles, Pythagoras (who combines mysticism and intellectualism) and finally,

Opposite above: Dancers of the Dogon tribe in Mali. The Dogon worshipped the companion star of Sirius long before its 'discovery' in 1830
Opposite below: 'Hitching post of the sun', scene of the Inca solstice ceremony in Peru

Shaman stilling the waves at Oumwaidjik on the Bering Strait. The heightened experience of the trance was a feature of ancient mystery religions

Socrates, Plato and Aristotle. The ancient religion of the earth and moon goddess has at last given way to the new solar religion of reason. In the section on Egypt, she emphasizes the continuity of the ancient death religion of the Neanderthals and the religious rites of the Egyptians – particularly the death and resurrection of Osiris, the god of the underworld who is son of the earth and sky. She also goes on to argue that in the religions of Central America – Aztec, Maya and others – this religion of death and resurrection had become savage and perverted, with the emphasis on blood, violence and sacrifice. Again, we should bear in mind that the Aztecs and Mayas also built pyramids oriented to the heavens, and went to immense lengths to work out an accurate calendar for purposes of religious ritual. (Peter Tompkins has written a sequel to *Secrets of the Great Pyramid* on the Mexican pyramids.)

It now becomes possible to see how an Egyptologist, writing about ancient Egyptian astronomy, can say: 'This somewhat casual and rather negative approach to the sky and "its affairs" suggests that to the ancient Egyptians they were of much less importance than ... terrestrial matters.... Throughout the three millennia of recorded Egyptian history we have nothing whatever to suggest that the movements of the Moon and the planets were systematically observed and recorded. . . .'[9] If we look at Egyptian astronomy from the 'scientific' point of view, it fades into insignificance. To grasp its importance, we have to know that, for the Egyptians, it was part of a religious mystery. But for the Egyptians, this religious knowledge *was* science, an attempt to grasp and understand the universe.

The same point emerges in Robert Temple's fascinating book *The Sirius Mystery*. Temple speaks of a modern African tribe in Mali called the Dogon, who possess a curious piece of astronomical knowledge: that the dog star, Sirius, is actually a double star. The Sirius we can see – the brightest star in the heavens – is now known as Sirius A. Circling around it is an invisible star we call Sirius B. Sirius B is a white dwarf – a tiny star of immense weight, whose atoms have collapsed in on themselves so that a piece the size of a pinhead weighs many tons. Sirius B was discovered by the astronomer Friedrich Wilhelm Bessel in the 1830s, when he observed curious perturbations in the orbit of Sirius, and reasoned that an invisible star must be tugging at it. The Dogon have a tradition about this invisible companion to Sirius, which they call the Digitaria star. They know that it is the 'smallest and heaviest of all stars', that it rotates on its axis, has an elliptical orbit, and revolves around Sirius A every fifty years. The Dogon regard Sirius B as the origin of all creation.

Temple argues that the knowledge of the Dogon about this invisible star can be traced back to ancient Egypt. He points out that a tribe related to the Dogon call Sirius B 'the eye star', and that Osiris, the companion of Isis (Sothis or Sirius) is represented in hieroglyphics with an eye. He interprets various Egyptian myths – of Anubis (the dog-headed god), Nephthys and Isis – as indicating that the Egyptians also knew that Sirius had an invisible companion. And he argues that the source of this strange knowledge could have been extra-terrestrial visitors from the star system of Sirius. But we have already mentioned the objection to the 'ancient astronaut' theory, as applied to early civilizations: if they obtained their astronomical knowledge from spacemen, why did they not also know that the sun is the centre of the solar system, and that the earth is a mere planet? We may argue that the knowledge was forgotten, as so much astronomical knowledge seems to have been forgotten in the millennia before Christ; but surely such a *simple* piece of information would have survived, if only as a tradition?

And what alternative theory can be offered to explain the knowledge of the Dogon – if we reject the notion that some wandering explorer took the trouble to impart this abstruse piece of information to them sometime after Bessel discovered it? Coincidence? That is not as absurd as it sounds. In *Gulliver's Travels*, published in 1726, Swift describes one of the discoveries of the *savants* of Laputa:

> They have likewise discovered two lesser stars, or satellites, which revolve about Mars, whereof the innermost is distant from the centre of the primary planet exactly three of his diameters, and the outermost, five; the former revolves in the space of ten hours, and the latter in twenty-one and a half; so that the squares of their periodical times are very near in the same proportion with the cubes of their distance from the centre of Mars. . . .

Swift was not quite correct. The innermost of the two satellites, Phobos, is one and a half, not three times, its diameter from the centre of Mars; and Deimos is about three and a half, not five diameters, away. Not bad, one might say, for the extremely poor telescopes of that period. . . . But the two satellites of Mars were not discovered until

1877 – by Asaph Hall – a century and a half after *Gulliver* appeared. Where the sidereal period (time of revolution) was concerned, Swift's guesses were even closer: Phobos *does* revolve in approximately ten hours, and Deimos in just over thirty.

How is such an incredible coincidence possible? Hapgood, who mentions the mystery, thinks that Swift may have heard some ancient tradition, perhaps dating back to antiquity; but Swift lived in eighteenth-century London, and neither there nor in Dublin was he likely to pick up such an odd piece of mythology. Like Dr Johnson, he was a classicist, and there is no such tradition in the classics.

Gulliver discoursing with the King of Laputa. In making fun of 'Laputian' astronomy,
Swift actually stumbled upon the fact that Mars has two satellites

Again, I would suggest that we are dealing with some intuitive knowledge originating in the 'dark side of the mind' – like the 'guesses' of Poe's *Eureka*. When a man of genius needs to invent some curious fact or theory for the sake of verisimilitude, he flings his mind open, as it were, and takes a lucky dip. And if that 'dark side of the mind' *has* access to 'distant facts', then it may well supply the correct information. We are speaking, in fact, of a form of what Jung calls synchronicity: coincidences that seem a little too complex to be pure coincidence. I can cite an example that occurred in the past day or two. I happened to find a copy of *Russian Nights* by the nineteenth-century writer Vladimir Fedorovich Odoevsky which I had lost for many years; reading the introduction, I came across the information that he had published his early works in a journal called *Mnemosyne*. It crossed my mind to wonder if this was the Greek god of memory.* The following morning I opened Gertrude Levy's *Gate of Horn*, to find on the first page a reference to an article on labyrinths in the Dutch journal *Mnemosyne*.

Hardly, I agree, a very exciting coincidence; but I shall probably not come across the name again for another ten years; and, as far as I know, I have not come across it in the past ten. I could cite other, more complex, examples – of a name heard for the first time, and encountered again – in totally different contexts *four* times within minutes.

The writer Jacques Vallée recently cited an amusing example of synchronicity. He had become interested in an obscure biblical prophet named Melchisedek, and was trying to find out all he could about him. The next day, he took a taxi, and asked the woman driver for a receipt; she signed it 'G. Melchisedek'. Wondering if it was a commoner name than he had supposed, he looked it up in the Los Angeles telephone directory. There was only one Melchisedek – his driver. Vallée said it was as if he had stuck up a note on some universal notice board, 'Wanted – information on Melchisedek', and some bumbling genie had said: 'How about this one?'

Vallée makes the interesting suggestion that our unconscious memory may work on the binary storage system of a computer rather than the serial system of a library – where books are arranged in alphabetical order; but this does not seem to me to explain the mystery. What seems to be involved here is something more like Jung's 'collective unconscious', some gigantic store of information that is accessible to the individual unconscious mind. Synchronicity, however, is not simply a matter of 'unlearned knowledge', but of synchronous *events* – as if you had asked a librarian to find other examples of the same thing, and she had dropped a book on your desk.

The logical conclusion – if we do not find these suggestions wholly ridiculous – is that the 'dark side of the mind' is somehow able to influence events, as well as gain access to hidden information (like whether the sixth number of a Fermat series is actually a prime). If this is so, we need a new word to supplement Jung's synchronicity. But in fact there is a perfectly good old word: magic.

We think of magic as an absurdity because it is 'outside science', and therefore unreasonable. The ancient Egyptians, Babylonians and Celts had a simpler and wider viewpoint: magic was part of a wider knowledge system that embraced science.

As if to underline my point about synchronicity, I received in the mail this morning a new book by the scientist Arthur M. Young, inventor of the helicopter, consisting of extracts from his journals for 1945–7. After reading a book about the psychic Edgar Cayce, Young comments:

What is our scientific civilization? Nothing? No, it is a very impressive something, but this something is really not what we thought it was. What it is, is the power of duplication. [That is, the scientist being able to duplicate an experiment in his laboratory.] Once upon a time there was magic. Civilization learned to reproduce, reduce to formula, and duplicate a part of this magic. That is our world of science. The rest of the magic has been tossed on the junk heap. . . . If we conscientiously seek out the final answers on, say, physics, it will be found that there is no evidence that denies unduplicatable magic.[10]

* In fact, it is the goddess.

Again, in a remarkable book called *The Origin of Consciousness in the Breakdown of the Bicameral Mind*, Dr Julian Jaynes puts forward the startling thesis that as recently as three thousand years ago our ancestors lacked all self-awareness – the power to look inside themselves and ask, 'What do I feel about this?' Jaynes argues that their attention was turned wholly outwards, towards the external world, and that they would have no idea of what we might mean by the phrase 'to withdraw into oneself'. Their decisions, Jaynes suggests, were made for them by the right half of the brain, which then communicated to them in the form of voices, which sounded inside their heads – like audial hallucinations.

It is impossible to accept his thesis in this simplistic form, even though the heroes

Isis, wife and sister of Osiris, with the sun's disc between her horns

of *The Iliad* do seem to take their orders from the gods. But Jaynes's impressive mass of argument does seem to demonstrate that 'ancient man' had a different kind of consciousness from ours, something more direct, intuitive, less self-divided. It is almost impossible for us to *think* our way back into this kind of consciousness, because thinking is a left-brain activity. But most of us have experienced it in states of deep relaxation or sudden delight – a sense of wider vistas of reality than we are normally accustomed to.

If Schwaller was correct about the Egyptians, then they represent a watershed in the history of civilization. Their knowledge was vast and far-reaching, but unsystematic – an intuitive hotchpotch. (Schwaller's wife, Isha Schwaller de Lubicz, has caught something of this in a novel called *Her-Bak*, a fictionalization of her husband's theories about ancient Egypt, which is clumsily written, but full of important insights.) Intuitive people have a natural dislike of systematization, which clamps them in a strait jacket. But the trouble with intuition – or doing things by rule of thumb – is that you have to keep on working out your basic premises every time you approach a new problem. I can, for example, easily work out the method of extracting a square root; but it is much simpler to have the *formula* in my head, even if this makes me lazy.

What is more, once people start to take the trouble to systematize their knowledge, to trap it in tables and formulae, they make an important discovery: that they can extend it merely by manipulating the formulae correctly. In the same way, once you have taken the trouble to learn the method for, say, making furniture – with mortice and tenon joints – you discover that it is far more efficient than the rule-of-thumb method using a hammer and a box of nails. It also looks tidier. So even the most naturally intuitive person must finally learn the advantages of being systematic.

This explains something that many people must have felt in museums containing ancient artefacts – of the Sumerians, Minoans, Hittites, and so on: how a 'highly advanced' civilization could apparently be contented with living conditions that were little better than those in an African village. Their religious and scientific knowledge were far in advance of their everyday life, which continued to be a matter of rule of thumb.

The ancient Egyptians appear to represent a turning point. Perhaps because their dependence on the Nile forced them into habits of regularity, they seem to have decided to make a tremendous attempt to systematize their lives. We can see this in their temples, which make an instant impression of order and regularity – the vast yet completely symmetrical columns, the perfect right angles, the almost invisible joins in the masonry. Yet the basically intuitive approach to knowledge remained; their geometry and astronomy were simply a sub-department of their religion. Then, towards the end of the third dynasty, some unknown high priest decided the time had come to combine all their knowledge into one gigantic system – a kind of ancient Egyptian *Summa Theologica*. To systematize astronomy, the Great Pyramid was constructed. And, without realizing it, the Egyptians had taken the decisive step into the modern world, the world of science and reason and the left-brain ego.

Because much of this knowledge was regarded as sacred, to be imparted only to initiates, it was lost during the next two thousand years – surviving only as a magical tradition about Hermes Trismegistos (or Thoth), the founder of magic and of writing. Much of the tradition became garbled and confused. But it made no difference. The new spirit had come into existence, and it would change the course of civilization – and, more important, of human evolution.

Atlas bearing a celestial globe on his shoulders

CHAPTER THREE

THE AGE OF ABSTRACTION

There is a simple trick involving numbers that can be guaranteed to produce astonishment at any party. You ask someone to write down his telephone number, then to write it a second time with the figures jumbled up. Next, tell him to subtract the smaller from the larger number, and keep on adding up the figures in the answer until he has reduced it to one figure. (So 19 becomes 10, which in turn becomes 1 plus 0 – that is, 1.) When he has finished, you may tell him authoritatively: 'The answer is nine.'

You can afford to be dogmatic; for the answer is *always* nine. It works with any set of figures, no matter how small or how large. Jumble up the figures, subtract one from the other, and the answer always reduces to 9.

I have no idea why this is so, and have never come across a mathematician who could explain it. It is just one of those peculiar properties of numbers.

The first man to become aware of the strangeness of numbers was the Greek philosopher Pythagoras, born on the island of Samos in about 570 BC. He seems to have been a man of extraordinary charisma for there are dozens of legends about him, and he is reputed to be the founder of magic. He was not, of course – magic was thousands of years older than the Greeks. In fact, Pythagoras was the exact opposite of a magician; he is the first important historical representative of anti-magic. That is, Pythagoras's vision of the universe is seen from the point of view of the left brain.

The career of Pythagoras need not concern us here; it is enough to say that, according to legend, he was sent to Egypt by the tyrant Polycrates of Samos to study the Egyptian mysteries. Diogenes Laertius tells us that he associated with the Chaldeans and the Magi. He had already become an adept in the Orphic mysteries of Greece. Returning to Samos, he discovered that his patron had changed for the worse and become a dictator, so he left for Crotona in southern Italy – which had been colonized by Greeks. There he spent most of his life, building up a reputation which spread all over the Mediterranean world, and finally arousing so much popular jealousy and resentment that he was forced to leave. (Diogenes says he

Pythagoras, founder of the rational intellectual tradition of the West. From Chartres Cathedral

was burned to death by the mob, but it seems fairly certain that he escaped and died, at the age of eighty, in Metapontum.)

By training, Pythagoras was a mystic; by inclination, a scientist. Mystics do not found secret brotherhoods, as Pythagoras did; nor do they preach that political action is an important part of philosophy. They follow the road of inner vision. Pythagoras spent his life studying the universe around him and trying to understand it in terms of numbers. His starting point seems to have been the insight that the pitch of a note depends on the length of the plucked string. The legend declares that Pythagoras was passing a smithy, and was intrigued to hear that the four blacksmiths made different notes as they struck the anvil with their hammers. He investigated, and discovered the hammers were of different weights. He went home, suspended a similar range of weights on identical lengths of string – producing different degrees of tautness – and discovered that the strings now gave off the same series of notes as the hammers. And when the strings were shortened in proportion to their weights, they again produced the same series of notes. . . . The only objection to this story is that *all* hammers would produce the same note when striking the same anvil; but we have only to change the four hammers into four anvils – of different sizes and weights – to have a perfectly serviceable legend.

Pythagoras was excited at the idea of associating measurements of length, a quantitative experience, with musical notes, a qualitative experience; he suspected that he had stumbled on some basic secret of the universe. He went on to theorize that the distance between the planets also corresponded to musical notes, so that in moving around the earth, they actually made a kind of music – the 'music of the spheres'. Only the mystic, said Pythagoras, or the lover of wisdom, can hear this music. . . .

The first thing that strikes one about this concept is that it *is* an idea, an interesting abstraction, which makes no obvious appeal to our intuition. This is characteristic of so many of Pythagoras's ideas; they are the creations of an incorrigible intellectual rather than a mystic. This intellectualism is particularly obvious in his thoughts on mathematics. He was fascinated by out-of-the-way pieces of information about figures. He once defined a friend as 'the other "I", like 220 and 284'. This referred to the fact that 220 can be divided by 1, 2, 4, 5, 10, 11, 20, 44, 55 and 110, and that these add up to 284, while 284 can be divided by 1, 2, 4, 71, and 142, and these add up to 220; so 220 and 284 are two 'amicable numbers'.

He also noted that 'square numbers' (like 4, 9 and so on) can be formed by adding on successive odd numbers, so 4 plus 5 equals 9, 9 plus 7 equals 16, 16 plus 9 equals 25. No one would deny that such observations are interesting; but are they *important*, as Pythagoras believed? The other night, lying awake, I tried to recall the above theory about squares; instead, I stumbled upon the realization that the difference between any two successive squares – say 5 squared (25) and 6 squared (36) – is the sum of the two numbers themselves – 11. In the middle of the night, this struck me as an interesting observation; and, since I lack a mathematical turn, I felt rather pleased with myself. But the next morning, I saw that this 'discovery' becomes *self-evident* if you draw a square number as an actual square divided into smaller squares. It then struck me that this same objection applies to all theories about numbers; if your powers of visualization were strong enough, they would all become self-evident. (This is what Wittgenstein meant when he said that mathematical statements are tautologies.) Which conclusively destroys the foundation of Pythagoras's number mysticism. For the flat truth is that the realm of numbers tells you nothing whatever about the realm of actuality. To know that 220 and 284 are 'amicable' numbers does not really add anything to my knowledge of friendship. Pythagoras's number mysticism was, quite simply, a mistake, a fallacy. So was his belief that numbers and ideas are the 'true essence' of things, and that material objects are only crude copies of this greater reality. So, too, was his belief in the music of the spheres. In fact, that great founder of the modern intellectual tradition was a charlatan – unconscious, perhaps, but a charlatan nevertheless. He had blinded himself with abstractions because he had lost touch with the basic realities – the earth under his feet and the planets overhead.

But how did he come to lose touch? We can only suppose that the evolution of scientific insight, from the days when Khufu built the Great Pyramid, had produced a new, alienated sensibility. The 'split' posited by Julian Jaynes had occurred; the human mind had become 'bicameral'. Mathematics, like language, is basically a left-brain function. And the left cerebral hemisphere is our practical side – Martha, as compared to the right brain's Mary. With the coming of Pythagoras, Martha became the boss of the household.

It is important to understand just how this came about. Language and arithmetical calculation are *automatic* functions; we have learned them by rote. The part of the brain that deals with such activities – the 'automatic pilot' or robot – is the cerebellum, a part of the 'old brain' that lies underneath the cerebral hemispheres. When you are engaged in *creative* thinking, the right and left brains collaborate closely. But if you are forced to talk a great deal, or add up whole columns of figures, the right brain begins to feel itself superfluous, and the left brain and the automatic pilot form a new alliance – like two schoolgirls deciding that a third member of the group is rather a bore. When this happens, they create what I can only call a 'false self'. Everyone has had this experience – for example, after watching television for too long or reading too much. We begin to feel alienated from our own lives; the world around us begins to seem oddly meaningless, and we find it hard to distinguish between something that has actually happened and something we have only read about or dreamed. This 'false self' looks at the world from *outside*, as if looking in through a shop window; it does not feel involved in the world. I have discussed the phenomenon at length in a book called *The Outsider*, in which the thesis is illustrated mainly by examples drawn from the nineteenth century, the age of romanticism and alienation. But the phenomenon can be traced back to Pythagoras. We learn, for example, that when a certain Leontius asked him what art he practised, Pythagoras replied that life could be compared to the Olympic Games, where there were three main types of people: the athletes, who sought glory, the stallholders, who wanted to turn an honest penny, and the spectators. This third group, he maintained, was the most noble of the three, for it sought neither applause nor cash, but only to contemplate the wonderful spectacle. Which, said Pythagoras, is why he himself was a spectator, a philosopher. (He invented the word philosophy.) This notion of being a spectator rather than a participant is typically romantic, and it also reveals the alienation of the intellectual from the real business of living. (It was not until the nineteenth century that a philosopher, Johann Gottlieb Fichte, had the good sense to state that a man is not truly 'himself' until he hurls himself into action.)

Pythagoras taught that the body is a tomb, and that man has to triumph over it. To do this, he has to achieve a state of ecstasy. (Oddly enough, Pythagoras took a poor view of sex, declaring that it makes a man 'less than himself'.) Here again we can hear the alienated intellectual for whom life is a gigantic question mark because he sees it from the point of view of the 'false self'.

Ironically, the Pythagorean ideas suffered their greatest blow through one of the master's most interesting discoveries – the so-called irrational numbers. The ratio of the diameter of a circle to its circumference is $3\frac{1}{7}$. But if you try to turn this into decimals, it is impossible; the decimal for one-seventh begins .142857, and then repeats itself an infinite number of times. So Pythagoras's belief that the universe is built out of neat whole numbers was exploded. Legend has it that the Pythagoreans were shattered by this discovery, and decided to keep it a secret. One disciple, Hippasus, let it slip and, according to Proclus, the gods punished him by causing him to be drowned in a shipwreck. A more plausible tradition says that he was put to death by his fellow disciples. At all events, the revelation of irrational numbers caused dismay among the mystical brotherhood, and eventually seems to have led to its

*Early Greek concept of the universe, showing a disc-like earth
floating on water inside a sphere*

dissolution. Nevertheless, its influence remained powerful. And by fortunate or unfortunate chance, it had a powerful fascination for a man who exercised immense influence on European ideas: Plato. Plato actually elaborates the doctrine of a 'musical universe' in the *Timaeus*; and his most famous doctrine – that 'ideas' are somehow more real than material objects – comes straight out of Pythagoras.

Now where astronomy was concerned, Pythagoras was not distinguished for any brilliant insights. He knew that the earth was round, but thought that it was at the centre of the universe, and that everything revolved around it. He is hardly to be blamed for this, of course, since all other philosophers believed the same thing. His predecessor, Thales, believed that the earth is a kind of saucer floating on water and that everything in the universe is made of water. Another predecessor, Anaximander, produced the astonishing idea that the sky is a thick skin, like the bark of a tree, and that the sun, moon and stars are not really solid objects at all, but a blazing fire that burns on the other side of the skin. He thought the earth was shaped like a cylinder. His contemporary and friend, Anaximenes, thought the sky was like a glass bowl, with the stars stuck like luminous nails on the inside.

The universe according to Philolaus. The central fire is orbited by a counter-earth, the earth, moon, sun, planets, and, outermost, the stars

Pythagoras at least convinced everybody that the stars and planets were real, with the result that the philosophers of the following century (he died around 500 BC) thought up ingenious mechanical models of the universe – an important step in the direction of scientific astronomy. Philolaus took the immensely important step of suggesting that the earth moved. Yet, oddly enough, he did not suggest that it rotated on its axis, but that it moved around in a small circle, like a horse on a child's roundabout.

Philolaus believed that the point around which the earth revolves is a 'central fire'. Yet this fire is not the sun – which was regarded as one of the planets; it is a kind of 'hearth of the universe', and we are unable to see it because the earth revolves with its back to it. It seems strange, under the circumstances, that Philolaus should have posited this 'central fire'; again it looks as if, like Swift and Poe, he had received a glimpse of some basic truth about the universe, but somehow misinterpreted it with his intellect.

Another of Pythagoras's disciples, Ecphantus, advanced the supposition that the earth rotated on its axis, and that this explained the apparent revolution of the heavens. So did Heraclides of Pontus, a disciple of Plato. But Heraclides encountered a problem that was to baffle astronomers for the next two thousand years; the planets. Seen from the point of view of the earth, the planets have a very peculiar habit. Instead of revolving around the earth in straightforward orbits, they 'wander' – the word planet means 'a wanderer'. Anyone who has ever read a book on astrology will know that the planets do not always proceed in a forward direction; they also have short periods when they go backwards. Of course, this motion is only apparent; it results from the fact that the earth is also moving and periodically 'passes' other planets. And, as when one train passes another, the slower appears to be going backwards.

If all the heavenly bodies are really stationary, and only the earth is moving, why should the planets 'wander'? The problem looks insoluble in terms of an earth-centred universe; but Heraclides came up with a brilliant solution. We can see that the sun travels around the earth. But Venus and Mercury always seem to be close to the sun, which argues that they may be its satellites. And if we suppose that the sun journeys round the earth while the planets journey round the sun, then it can be seen that the orbits of the planets would be shaped like a corkscrew. Therefore we would sometimes see them coming towards us and sometimes moving away from us. . . .

Heraclides died in 310 BC. The golden age of Athens was already drawing to a close; its place was taken by Alexandria, the city on the Nile founded by Alexander the Great. Alexander of Macedon began his conquests in 334 BC, and died of a fever a mere eleven years later, at the age of thirty-three. But he had already chosen the site of the city that was to be named after him. It was built by one of his generals, Ptolemy, who also founded the famous dynasty. Ptolemy, like Alexander, had been a pupil of Aristotle, and he loved learning. He built the great library and museum of Alexandria, and made it known that scholars would be welcome. As a result, Alexandria replaced Athens as the cultural centre of the Mediterranean.

One of the scientists who moved to Alexandria was Aristarchus of Samos, a Pythagorean. He had inherited from Pythagoras the notion that the earth is a globe (which Pythagoras may have learned from the Egyptian priests). Now he made the suggestion that the earth revolves around the sun, while it also rotates on its axis. This notion seems to have aroused little interest, and the treatise in which he originally suggested it has vanished. No doubt his contemporaries thought the idea less satisfying than Plato's suggestion that each of the planets is attached to a kind of crystal sphere. . . . But the ancients *were* impressed by Aristarchus's treatise on the distance of the moon and sun – rightly so, since his method of determining these distances was one of the most remarkable achievements of ancient science.

The question of the distance of the heavenly bodies had always intrigued astronomers, but it seemed unanswerable – until the Greeks invented trigonometry, the

Hipparchus watching the heavens at Alexandria. A nineteenth-century engraving

science of right-angled triangles. From Pythagoras's discovery about the sides of a right-angled triangle, they went on to recognize that if you know the size of the angles, you also know the *proportion* of the sides to one another. Aristarchus reasoned that if he could observe the sun and moon when they create a right-angled triangle with the earth, he would know their relative distances. He knew that the sun is further away than the moon, because the moon sometimes eclipses it. He also knew that the moon shines by the reflected light of the sun. He therefore reasoned that when we see exactly half the moon, this must be because its profile is turned towards the sun:

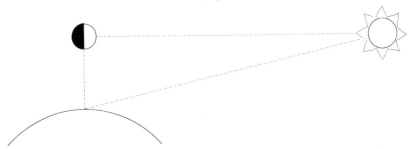

All he had to do was to wait until the sun and moon were in the sky at the same time, and the moon showed only half its face. Then by measuring the angle from himself to the sun, he could construct a right-angled triangle. The angle he measured was 87° (so the triangle is far longer and narrower than shown in the diagram). Simple trigonometry reveals that the tangent of 87° is 19.081; so the sun, according to Aristarchus, is about twenty times further away than the moon. His measurement was at fault, of course, for the actual figure is closer to 360, and the angle in question should have been almost 90°. Aristarchus's error was caused by the difficulty of judging when the moon is precisely half full.

Since the moon neatly covers the sun at eclipses – appearing to be just the same size – Aristarchus judged that the sun must be twenty times larger than the moon. From the size of the earth's shadow on the moon during a lunar eclipse, he also worked out that the earth's diameter is about three times that of the moon – an astonishingly accurate figure. More calculation convinced him that the distance of the moon from the earth is about fifty-six of the earth's radiuses – again, an incredibly accurate figure, coming close to the actual distance of a quarter of a million miles.[1]

Of course, Aristarchus did not know the size of the earth's radius – Eratosthenes was only a boy at the time he made his measurements. But it was perfectly obvious – from the experience of travellers and seamen – that it ran to hundreds, possibly thousands, of miles. What he *could* quite easily calculate was that if the earth's radius is a thousand miles (a mere quarter of the true figure), then the moon is fifty-six thousand miles away, and the sun well over a million.

To the modern mind, accustomed to immense numbers, these figures sound unspectacular. But it is necessary to make the effort of imagination, and try to imagine their impact on the intelligentsia of Alexandria. Their first reaction must have been incredulity; then, when they saw that Aristarchus's calculations were free of error, stunned amazement. They thought of the earth as the largest object in space, the centre of the universe, with a relatively small sun and moon circling around it at a distance of a couple of miles. The story of Daedalus told them that if a man with wings flew too high – say half a mile – the sun would melt the wax that held his wings together. If Aristarchus was correct – and his treatise made him one of the most celebrated astronomers of the Mediterranean world – the sun's distance was so immense that a man could fly a hundred miles high, a thousand, and still not get much warmer. It must have produced a kind of intellectual vertigo.

Aristarchus went on to draw the natural conclusion from his calculation of the relative size of the sun. If it is seven times larger than the earth, then why should it travel round it? It is natural to suppose that it should be the other way round.

This confronts us with an interesting question. If Aristarchus's theory about the sun and moon made him the most respected astronomer of his time, then why did his 'Copernican theory' make so little impact? Surely the astronomers of Alexandria must have *seen* that it made sense for the smaller body to revolve round the larger one, and not *vice versa*? And, moreover, that this would explain the mystery of why the planets sometimes seem to go backwards? In his *History of Astronomy*, J. L. E. Dreyer suggests that perhaps Aristarchus 'had merely thrown out this suggestion or hypothesis without devoting a book or essay to it'.[2] But this is contradicted by Archimedes, who says that Aristarchus 'had published in outline certain hypotheses'.[3] Could it be that the 'outline' was so brief that no one paid much attention to it? Hardly, since Plutarch was still discussing it more than three centuries later.

The answer to this question, I would suggest, is far more important than it looks – so important, indeed, that it throws a new light on man's intellectual development over the past three thousand years. And in order to grasp it, we must cast our minds back to the comfortable world picture of the Egyptians and Babylonians, who took it for granted that our earth is the centre of the universe, and that the sun god drives his chariot round it once a day to provide the human race with light and heat. This idea was wrong; yet it confirmed their natural intuition that there is a close relation between man and the earth and the stars. They had no fear of knowledge, because they believed it came from the gods, who had given man a glimpse of their intentions. The Egyptian pharaoh was, after all, himself a god. . . .

Sometime between the building of the Great Pyramid and the birth of Pythagoras two thousand years later, an immense change took place, a change that virtually made man into a different kind of creature. Julian Jaynes calls it the coming of the 'bicameral mind'. Man developed 'head consciousness' instead of the old, unified intuition of nature. But why did this change come about? And how? Was it a gradual process, resulting from the complexities of Bronze Age civilization? Or was there a sudden crisis which caused it to happen overnight?

The answer is probably: both. Jaynes believes that the 'weakening' of the old unified consciousness began about 2500 BC, with the widespread use of writing; I have suggested that the building of the Great Pyramid, at the same date, marks a watershed in human history.

We now know that immense and catastrophic changes took place all over the Mediterranean more than a thousand years later, plunging Bronze Age civilization into chaos. To begin with, there was the tremendous volcanic explosion on the island of Santorini that took place in about 1500 BC. Santorini is a small island in the Aegean, about sixty miles north of Crete – where the Minoans had developed their own remarkable civilization. We can judge the devastation that occurred from our knowledge of what happened during the eruption of Krakatoa in 1883. When the great volcano in the Sunda Strait had finished erupting, it collapsed below sea level; the water rushed in, and there was an explosion that caused a hundred-foot tidal wave, which swept over islands, destroying whole towns and killing thirty-six thousand people. The shock was felt as far away as Paris. The explosion of Santorini is estimated to have been three times as great as that of Krakatoa, and in an inland sea like the Mediterranean the damage must have been inestimable. The palace of Knossos, in northern Crete, was destroyed, and the Greek islands must have been swept clean of all life.

Yet this, it seems, was only the beginning. Archeological evidence reveals that the palace of Phaistos, in southern Crete, was also destroyed at about the same time; and even the largest tidal wave could not have passed over the intervening mountain range. The devastation of the whole Mediterranean area between 1250 and 1150 BC was caused by man, not by nature. An Egyptian text of the period refers to barbarian invaders called 'peoples from the sea'. Recent investigation[4] has shown that the 'sea peoples' were not a single racial group, but several different groups of nomads, pirates and footloose mercenaries. For various reasons, Mediterranean civilization was nearing the point of collapse in about 1350 BC – the devastation of Santorini, political and economic exhaustion, corruption, and squabbles among the small independent states. The 'sea peoples' attacked the bleeding civilization like sharks, and between 1250 and 1150 there was a catastrophic collapse into barbarism. It was so shattering that, in spite of the lack of written historical records, the memory was still alive seven hundred years later, when Thucydides spoke of it.

When civilization is plunged into a dark age, the survivors are those who possess most 'left-brain' qualities: efficiency, ruthlessness, a tendency to concentrate upon the present moment, nervous energy, vigilance (and the mistrust that goes with it), split-second reactions. This left-brain consciousness also generates nervous tension and impatience, which in turn generate cruelty. Man no longer feels a sense of

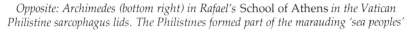

Opposite: Archimedes (bottom right) in Rafael's School of Athens *in the Vatican Philistine sarcophagus lids. The Philistines formed part of the marauding 'sea peoples'*

oneness with the universe, and the bovine serenity that goes with it. Trapped in the narrowness of the present, he no longer feels a sense of communion with the gods. Jaynes produces an interesting piece of evidence to support this thesis. Around the year 1230 BC, the Assyrian king, Tukulti-Nunurti, had a stone altar carved, with a picture of himself kneeling in front of an empty throne, the throne of the god. No king had ever before been shown kneeling – Hammurabi was always standing and listening to the god. Now the throne is empty; the gods have vanished; and the King, a suppliant and no longer a comrade, kneels in front of the empty throne. A cuneiform tablet from this period has the curious lines:

> One who has no god, as he walks along the street,
> Headache envelopes him like a garment. . . .

Headache, that familiar companion of nervous tension, haunts the man who no longer has contact with the gods.

A century later, another Assyrian king, Tiglath-Pileser, no longer follows the ancient custom of hyphenating his own name with that of the god, and the clay tablet

The bas-relief of Tukulti-Nunurti kneeling before the empty throne

from which we know of his exploits displays a new, shrill egoism and boastfulness. Moreover, his boasts are about a cruelty which seems to be new in history. His bas-reliefs in the British Museum show people being murdered in various ways – stabbed, decapitated, spitted on stakes. His soldiers overran towns and slaughtered civilians. His laws exacted the most bloody and violent penalties, even for minor offences. This is the cruelty of a man who is trapped inside the circle of his own personality – for whom other people are somehow unreal.

It is important to understand the distinction between the 'old' type of consciousness and the new. Jaynes's assumption seems to be fundamentally false: that 'ancient man' possessed no ego, no sense of 'I', and that instead of 'communing with himself' to make important decisions, he simply obeyed voices that came from the other side of his brain. This is physiologically unlikely, since the commissure joining the two halves of the brain certainly existed as recently as 2000 BC. But it also contradicts our everyday experience. When we heave a sigh of relief and relax deeply – perhaps stimulated by music or poetry, or simply a glass of wine – we do not lose touch with the right brain. On the contrary, we become more deeply aware of it. I lose the usual sense of being trapped in the present, the sense of ego. At the same time, memory works better; my own past becomes accessible to me. An odd feeling of confidence rises in me; I cease to feel a generalized mistrust of the universe and of fate. I become more 'intuitive'; and people who have experienced flashes of 'second sight' assert that they are more likely to occur in these moods of relaxation. In short, Jaynes's hypothesis about 'voices' from the right brain is quite unnecessary. The 'voices' are those of my intuition, and do not express themselves in words but in 'promptings'.

There is one great disadvantage to these moods of 'intuitive consciousness'. They make us lazy. If, for example, I have an important insight in one of these relaxed moods, it costs me the utmost effort to force myself to pick up a pencil and write it down. I feel sure it will still be there the following morning – and it never is. Intuitive consciousness dislikes the favourite activities of the left brain: thinking, writing, calculating. (This is why most children hate mathematics.) So to be 'marooned' in left-brain consciousness is by no means an unmitigated evil. It may reduce man to a mere creature, kneeling in front of an empty throne. But it also confers on him certain godlike powers. Once he commits himself to this thinner and poorer type of consciousness, and learns to peer at the world through a kind of microscope, once he starts creating lists and tables and mathematical formulae, his power over nature increases a hundredfold. Intuitive insights tend to vanish a few hours later; but once something has been written down, it is there for all time. Moreover, when knowledge is committed to writing, instead of being passed on by word of mouth in sacred schools and remote temples, it becomes available to everyone who happens to be born with an appetite for it. There is strong reason to believe that Egyptian civilization collapsed because it was *too* hierarchical; if you were born an artisan you stayed an artisan. But Socrates was the son of a midwife; and the philosopher Bion boasted that he was the son of a freedman who used to wipe his mouth on his sleeve. The real difference between Athens and ancient Thebes is that the Greek mind was 'bicameral'.

It took many centuries, however, for man to recognize the advantages of 'head consciousness'. To begin with, it only made life intolerable; men were bad tempered and cruel. They continued to hanker after the old gods and the old unified consciousness. The great religion of Dionysus, which swept down from Thrace in the seventh century BC, was a revolt against head consciousness, a craving for ecstasy and violence. (It is no accident that Dionysus is also the god of wine.) The guardians of Greek religion sensibly decided not to oppose it, and made Dionysus one of the gods of their pantheon. More than two and a half millennia later, Nietzsche would declare himself a 'disciple of the god Dionysus', and denounce Socrates for glorifying 'head consciousness'. The modern psychedelic cults reveal that we still hanker after the ancient consciousness we began to lose when the Egyptians laid the first stone of the Great Pyramid.

It now becomes possible to see why the Greeks paid no attention to Aristarchus's theory that the earth went round the sun. It was not mere lack of interest in a wild idea, but something deeper. Aristarchus had already turned their universe upside down by revealing that the sun is millions of miles away. But to propose that the earth was a mere satellite of the sun was more deeply disturbing. One contemporary, Cleanthes, suggested that Aristarchus should be censured for impiety towards the gods – and we know from the fate of Socrates what that could mean. The Greeks had been dragged far enough out of their comfortable womb of intuitive consciousness; this idea of Aristarchus was frightening, like some story of inhuman brutality. Pascal expressed the same feeling two thousand years later when he wrote, 'The eternal silence of these infinite spaces terrifies me.' The Greeks ignored Aristarchus's theory because they could not absorb it. Whether it was true was beside the point; they experienced a profound revulsion at the very idea.

This insight also tells us a great deal about the intellectual history of the next two thousand years. Man continued to hanker after the old intuitive consciousness, as represented by the 'supernatural'. Plutarch – who lived in the first century AD – has a nostalgic little essay on the decline of the oracles. The ancient Greeks, like the Egyptians and the Chinese, believed that sacred forces reside in the earth, and that certain places – like Delphi – form a point of contact between men and the gods. Now that Greece is devastated and depopulated, says Plutarch, the gods have abandoned their sacred places, and no longer speak to men through oracles and sybils. Yet he continues to believe that nature is full of divine forces – which the Greeks called daimons – which influence human destiny; and the moon is connected with these daimons. We might add, by way of commentary, that Plutarch's remote ancestors had been aware of these divine forces through direct experience, while he knew them only by hearsay.

We know from the Bible that this nostalgia for the gods, for the days when man heard the voice of Jehovah, found expression in the expectation of a messiah, who would overthrow modern wickedness, and lead the people back to the simpler religion of their ancestors. The intensity of this longing explains the amazing story of the Christian Church and its conquest of the world. Greek civilization had collapsed, and the Mediterranean world was dominated by the Romans – the most determinedly materialistic and 'left-brain' civilization yet to appear. The gods were dead or had lost their power – but the craving for the old, unified consciousness was stronger than ever. Christianity was a primitive form of revivalism. It told people that they had immortal souls, and that they could get to heaven through suffering. It appealed to the feeling that the world had changed for the worse by announcing that it would shortly come to an end anyway. Jesus told his disciples that he would return *within the lifetime* of people then alive, and that the Day of Judgment would follow. In a sense, therefore, Christianity was an immense attempt to turn back the clock, an answer to the nostalgia expressed by Plutarch.

This also explains why the Christian Church remained so profoundly suspicious of learning (unless accompanied by endless protestations of orthodoxy). Without an understanding of the history of the 'bicameral mind', we shall be baffled by the endless and bloody squabbles about orthodoxy that divided the Church from the moment St Paul was in his grave. Orthodoxy was important because it was the logical alternative to the kind of free speculation that had produced this profound psychological craving for certainty. Christianity offered a relief from divided consciousness; it offered to reunite the bicameral mind through belief in Christ, and the salvation he had brought through his death on the cross. Once again man could act and believe with every particle of his being; he had been saved from his own scepticism by the Truth. The Truth would make him free – free from self-division, free from the tendency to ask questions, free from the fear of the 'infinite spaces' conjured up by the infidel philosophers. Now man had been given the truth, there was no need to speculate; speculation was the devil's favourite device for undermining faith. This is why the scientist, and Franciscan monk, Roger Bacon was to spend most of the last fifteen years of his life in prison; he advocated that science should be based on experiment and observation, not on the authority of Aristotle (who, like Dionysus, had been comfortably absorbed by the guardians of orthodoxy). But we are getting ahead of our story.

The Chaos of the Elements. An engraving from Robert Fludd's Utriusque Cosmi Historia, *1617.*
Mystery and messianic religions sought to restore the old intuitive consciousness

After the death of Aristarchus, some time around 250 BC, the school of Alexandria seems to have undergone a revolution similar to the one that occurred in Egypt with the building of the Great Pyramid: that is, its astronomers – men like Apollonius of Perga and Hipparchus – decided that the time for vague theories and generalizations was over; it was time for measurement and observation. Their observations only made it clear that Plato's theory that the planets were attached to crystal spheres was a failure. They therefore began to toy with a notion that had its origin in the 'corkscrew' theory of Heraclides. If the planets not only revolved round the earth in a circle, but also swooped around in small circles as they did so, this would explain why they sometimes went backwards. The system can be compared to a roundabout at a funfair on which chairs also whirl around on pivots. The small circles were called epicycles.

The library of Alexandria was burned down in 47 BC, when Julius Caesar was trying to subdue the city, and many scientific works perished – this is probably when

Fourteenth-century Florentine panel depicting Ptolemy and Astronomy

Aristarchus's essay disappeared. So we know less than we might about the theories of Apollonius and Hipparchus. But this makes little difference, since they were incorporated into the work of the man whose name became a synonym for astronomy, Claudius Ptolemy. In retrospect, Ptolemy's work seems an immense monument to wrongheadedness. He assumed that the earth was the centre of the universe and that it was stationary; all the other heavenly bodies revolved around it. The theory of epicycles explained why the planets sometimes stood still or went backwards – because they revolve in a small circle once a year as well as in their orbit around the earth. At a certain moment each planet would be travelling anti-clockwise round its epicycle, in the opposite direction to its motion in its larger orbit; so it appeared to stand still. Mercury and Venus travelled round the earth lined up with the sun, which is why they always appeared close to it. To explain other anomalies, Ptolemy supposed that the centre of their orbit was not the earth itself, but some point in space near the earth; so from the point of view of the earth, they moved in irregular circles, rather like a gramophone record with the hole off-centre. It was an absurdly complicated system, but it seemed to work. King Alphonso the Wise of Castile and León later remarked wryly: 'If the Lord Almighty had consulted me before embarking on the Creation, I should have recommended something simpler.' His instinct was correct. Nature usually *is* simple, and the Ptolemaic system (as it came to be called) is preposterous. Yet, like the 'science' of Aristotle, it came to dominate Christian Europe as the last word in scientific achievement. With Aristarchus, the history of ideas had reached a sticking point; with Ptolemy, the wheel began to revolve backwards.

The Roman Empire rose and fell; but the fundamentally materialist Roman civilization contributed little to astronomy, or to any science except engineering. If they studied the stars, it was only to foretell the future. Sailors *could* have used the stars to work out their precise position at sea, since Ptolemy had constructed a comprehensive star table; but they seem to have used the stars only to determine direction, not to establish their longitude. Ptolemy had also described the construction of an instrument for measuring the positions of the stars; it was called an astrolabe, and had a rotating dial on which degrees could be read off. The Greeks seem to have been formidably efficient with mechanics, an ability later forgotten. In 1900, Greek sponge divers off the island of Antikythera discovered a sunken wreck, dating from the first century BC, in which they found a small machine with geared wheels; it turned out to be an amazingly complex computer for working out the motions of the sun, moon and planets – probably for astrological purposes. Yet before this discovery, historians took it for granted that Greek knowledge of mechanics was crude and limited. It is an interesting lesson in how easily knowledge can be lost.

In the outlying island called Britain, Stonehenge was no longer used as a calculator of eclipses or a solar calendar; its purpose had also been forgotten. Yet in these remote islands there had been no violent catastrophe like the collapse of the Mediterranean world of 1200 BC. The Beaker people (who built the double bluestone circle of Stonehenge) gave way to the prosperous Wessex people, who arrived in about 1450 BC and traded with the Egyptians and Mycenaeans in the age before the great collapse. These wealthy merchants built the inner horseshoe of Stonehenge. Elsewhere in Britain, the original native population spread peacefully, cultivating the land, cremating their dead and burying them in urns (hence their name – the Urn people). In the sixth century BC, the Celts came over from Europe and ushered in the Iron Age. Their priests, the Druids, worshipped the fertility god in sacred groves. They probably made

Astronomicum Caesareum:

Historicus, diuina gerens, sophiæq; peri°,
Hic sua cognoscet, si bona nosse volet;
Namq; vetustatis mirator tempora rebus
Distribuet, verè dum canet historias,
Ipse sacri præses noctes æquare diebus

Discet, & hinc serie festa locare suæ;
Ipseq; naturæ rimator mira cometæ
Percipiet, nulli dicta vel acta prius;
Sed caueant animis adsint liuore perustis,
Hæc etenim labes cernere vera negat.

Ad Principem Thomam Thyrleleium Episcopū Vehreonasis vivitse

80

use of Stonehenge as a temple, and may even have performed human sacrifices on the altar (although their chief method of sacrifice was to burn the victims alive in wooden cages – the flames representing the sun).

The Celts chose ancient sacred sites – on hilltops – for their forts; and later, when England was converted to Christianity, these same places were again chosen as the sites for churches and monasteries – with the sun god transformed into St Michael. (The tower on top of Glastonbury is an example.) This is impressive evidence that, even though the earth forces were half forgotten, the ancient tradition lived on. In due course the Romans came, were horrified by the Druid practice of human sacrifice, and drove them into Wales. Then, in the fourth century AD, Rome was surrounded by enemies, and Britain was invaded by the Saxons. When the Britons sent for aid to Rome, the Emperor Honorius replied that they would have to fend for themselves. A remarkable general called Artorius held back the Saxon invaders for fifty years, defeating them in battle after battle, and becoming a legend in his lifetime; mortally wounded in battle by his nephew Mordred, he was carried to Glastonbury Abbey, and buried there in a secret grave in about 540 AD. Arthur became one of the great legendary heroes, a symbol of chivalry and civilization as the barbarians poured across Europe, and the Dark Ages closed in.

In 410 AD, only three years after the evacuation of Britain by the Roman garrison, Rome fell, sacked by the Visigoths. In proconsular Africa, the Bishop of Hippo, St Augustine, impressed by this fall of a mighty empire, began his book *The City of God*, in order to defend Christianity against the charge of weakening Rome by preaching love and compassion. He insisted that man turn away from the earthly 'city' towards heaven. Naturally, he had nothing but contempt for science, and he solemnly warned Christians against 'a certain vain desire and curiosity ... to make experiments ... cloaked under the name of learning and knowledge'.[5] His attitude is important because *The City of God* became the best-seller of the Middle Ages; it exists in more manuscript copies than any book except the Bible.

Shortly after 600 AD the prophet Mohammed founded Islam, and the newly united Arab nations set out to convert the world. Alexandria fell in 640 AD. The Arab leader Amr Ibn Al-'As is reputed to have pointed to the library and said: 'If these books agree with the Koran they are useless. If not, they are infidel. Burn them.' So once more, the library went up in flames.

Fortunately, this philistine attitude towards books did not last. In 765 AD, the great Caliph Al-Mansur, founder of the Abbasid dynasty, fell ill with a stomach complaint in his newly founded city of Baghdad. A hundred and fifty miles away, at a place called Jundi Shapur, there was a Christian monastery which had a famous medical school. The head monk went to Baghdad and cured the Caliph, who asked him to set up a hospital in the capital. Now it so happened that the monastery of Jundi Shapur was full of ancient Greek classics – some of them possibly from the library of Alexandria – and these included texts on astronomy. The Arabs had a practical reason for being interested in astronomy, since their mosques all had to face Mecca – and the magnetic compass had not yet been discovered. Al-Mansur ordered that the texts be translated into Arabic and, where necessary, up-dated. While this was happening, astronomy

Opposite: The Astronomicum Caesareum, *1540, by Petrus Apianus,*
a contemporary of Copernicus, who perfected the Ptolemaic world system.
The illustration symbolizes the eclipses of the sun and the moon

received another unexpected stimulus. A traveller from India arrived at the court of Al-Mansur in 773, bringing with him some Indian texts. Astronomy had reached India through Alexander the Great, and the traveller was carrying the Indian equivalent of Ptolemy's treatise, a book of star tables called the *Siddhartha*. Moreover, the Indian showed the Arabs a new way of writing figures, far less cumbersome than the Latin method: it placed the units in one column, the tens in the next, and so on – the method we still use today. It made mathematics much easier, and simplified the star tables.

The Arabs were fascinated by the new world of learning that had opened up for them – medicine, philosophy and astrology, as well as astronomy. For centuries, their

Arab astrologers making calculations. A European engraving, 1513

intellects had been starved in the harsh conditions of the desert; now, overnight, their world had changed. Baghdad became the eighth-century equivalent of Athens. Ptolemy's treatise on astronomy, the *Syntaxis*, was translated under the title of the *Almagest* (meaning 'the greatest work'). The caliph responsible for its translation was Al-Mamun, the son of Harun Al-Rashid, whom we have already met in connection with the Great Pyramid. Al-Mamun ordered his men to break into the pyramid because he believed there might be Egyptian star maps hidden there – clear evidence that some faint memory of Egyptian astronomy persisted down the centuries.

In the centuries after Al-Mansur, there were a number of great Arab astronomers: Alfraganus, Albumazar, Albategnius, Arzachel (to give the familiar latinized version of their names). Arab star names found their way into astronomy – like Betelgeuse, which derived from the Arabic *mankib al-gawzâ*, meaning the shoulder of Orion.

As for Europe, it had been overrun by successive waves of barbarian hordes – Visigoths, Huns (nomadic Mongols), Vandals (who pillaged Rome again), Burgundians, Ostrogoths, Saxons, Vikings. Spain had fallen to the Arabs as early as 715 AD, in the days before the Arab rediscovery of learning. Now the learning began to filter back into Europe again through Spain. The *Almagest* was translated into Latin, and caused a sensation which it is difficult for us to understand. Christian scholars (who were mostly churchmen) seemed to regard it as the revelation of a divine plan. And here we see, ironically, that Ptolemy's pedantry and lack of imagination were of some use after all. If he had been intelligent enough to see that Aristarchus was right, his book would have undoubtedly been banned by the Church as blasphemous. But since he believed the earth was the centre of the universe, there was nothing to take exception to.

Learning was back again. After centuries of intellectual stagnation, the appetite for it was ravenous. Man was never intended to underwork his brain; it produces a kind of moral dyspepsia. Ptolemy's system may have been cumbersome and absurd; but it was also complex and difficult, and that seemed to guarantee its genuineness. Besides,

Medieval astronomers with primitive thirteenth-century instruments

it was *new*, and the new has power to touch our springs of vitality. Ptolemy had the same kind of impact on the Middle Ages that Einstein was to have on the twentieth century.

What followed was a period that Arthur Koestler has called 'the thaw'. After 800 AD, Charlemagne began to push back the Arabs in Spain (although it would be many more centuries before they were completely evicted). The pride of Christendom re-awoke. Sea trade increased, as the triangular lateen sail replaced the old square sail, enabling ships to tack into the wind. The rudder made steering easier. Ships became larger. The magnetic compass was discovered. At Toledo, King Alphonso the Wise commissioned the most up-to-date star tables so far, and ordered that all his ships carry compasses. The Church invented the waterclock, and the mechanical clock soon followed. A new and more sophisticated plough revolutionized agriculture; so did the invention of the harness, which enabled farmers to attach horses to the plough. Universities were founded – Bologna, Paris, Oxford, Cambridge. The works of Aristotle were rediscovered, and caused the same kind of excitement as Ptolemy's had. As the Arabs were slowly forced out of Spain, Christendom went on the offensive with the Crusades; and one effect was that more Greek manuscripts from the Holy Land poured into Europe. In 1245, the young Roger Bacon was lecturing at the University of Paris on Aristotle, and conceived the idea of writing an encyclopedic work on all the sciences; his *Opus Majus* ranged from philosophy and theology to mathematics and optics. But his insistence on the importance of experiment and observation finally led the Church to silence him by sending him to prison. The ecclesiastical authorities felt the thaw had gone far enough; it had to be stopped.

This was easier to decide than to carry out; the appetite for knowledge was now too widespread. For a while, Aristotle himself acted as a kind of drag-anchor, for the scientists of the Middle Ages believed that every word he had said was true, and he had talked so much nonsense about science that he held up its progress for centuries. But in 1456 there occurred the event that inaugurated the new age of science. Johann Gutenberg invented the printing press. Now that it was no longer necessary to copy out books by hand information could be transmitted quickly and easily. One of the earliest printed books was a treatise pointing out that Aristotle and Ptolemy were in basic disagreement on some essential points of astronomy; it caused widespread dismay among the faithful. They would have been even more dismayed if they had realized that the authority of Aristotle and Ptolemy was about to be overthrown by one of their own number.

PART II

THE ERA OF
DISCOVERY

CHAPTER FOUR

THE HARMONY OF THE WORLD

The life of Nicolaus Copernicus is such a major watershed in European history that biographers find it hard to resist the temptation to dramatize. He is usually portrayed as a truly heroic figure, the complete Renaissance man. This is the kind of person he *should* have been, if history had been shaped by Cecil B. de Mille. The facts, unfortunately, are something of an anticlimax.

Copernicus was a timid and pedantic scholar, who spent his adult life trying to reconcile the universe of Ptolemy with the universe of Aristarchus, without having the courage or the imagination to discard Ptolemy and start from scratch. The motive that led him to question Ptolemy was not a basic dissatisfaction with the whole absurd system of epicycles and perfect circles, but a niggling worry about a fairly minor point in the Ptolemaic system – to do with the varying speed of the planets. Copernicus had an obsessively tidy mind – he was the type that Freud would label an anal erotic, meaning a fusser – and this inconsistency bothered him as a speck of dust bothers an orderly housewife. It led him to pull the Ptolemaic system apart and piece it together again; still with as many preposterous epicycles (in fact, even more), but with the sun at the centre of the system.

He was prevailed on to publish this idea towards the end of his life, and received the first copy of his book *De Revolutionibus Orbicum Coelestium* (*On the Revolutions of the Heavenly Spheres*) on his deathbed. It would be pleasant – and appropriate – to add that the book had the effect of a bombshell, and marked a turning point in the history of science. In fact, hardly anyone noticed. It would be another half-century before Kepler and Galileo leapt to the correct solution of the problem of the planetary orbits. Altogether, Copernicus must be one of the most unsatisfactory heroes in the history of ideas.

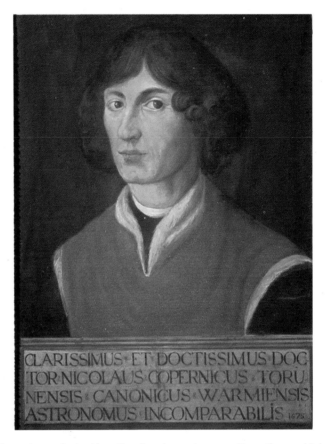

CLARISSIMUS ET DOCTISSIMUS DOC
TOR NICOLAUS COPERNICUS TORU
NENSIS CANONICUS WARMIENSIS
ASTRONOMUS INCOMPARABILIS 1675

Nicolaus Copernicus, reluctant begetter of modern astronomy. From Cracow University, 1575
Opposite: The alchemist's kitchen at the Collegium Maius, Old University, Cracow

He was born at Toruń, on the Vistula, on the border between Prussia and Poland, on 19 February 1473. His father was a rich merchant called Niklas Koppernigk. (In Franconian dialect *koepperneksch* still means 'a far-fetched, absurd proposition'.) We shall refer to the son by the latinized version of his name. When his father died in 1484, Copernicus (together with a brother and two sisters) passed into the guardianship of his uncle Lucas Watzelrode, who later became a bishop. Uncle Lucas seems to have been genuinely fond of his nephew, who was as submissive and self-effacing as his guardian was domineering and energetic.

At the age of eighteen, in 1491, Copernicus went to study at the University of Cracow. It was probably in Cracow that he became interested in astronomy – the university possessed a celebrated professor of mathematics and astronomy called Albert Brudzevsky. After four years of study, he returned to Toruń for two years, where his uncle was able to grant him a sinecure as a canon of Frauenburg Cathedral; as soon as he had been safely appointed, Copernicus went off to Italy to further his studies. In fact, he spent the next ten years at the universities of Bologna and Padua, taking his degree at the University of Ferrara. The reason for this latter choice seems typical of Copernicus's character: degrees at Ferrara could be obtained more cheaply

than elsewhere (it was a small university), and by slipping away to an obscure university, he was able to avoid the lavish hospitality which a newly promoted doctor was expected to provide for his friends, teachers and fellow students.

Back home, at the age of thirty-three, Copernicus became secretary and physician to his uncle the Bishop – he seems to have studied medicine at Padua. The Bishop lived in Heilsberg Castle, and was virtual ruler of the small principality of Ermland (or Varmia), right in the midst of the territory of the Teutonic Knights. The Bishop had no love for the Knights – a crusading order which had conquered heathen Prussia and wiped out most of its population. Since 1410, the Teutonic Knights had been engaged in more or less constant war with Poland, and had been compelled to hand over this large chunk of their territory – Ermland – in 1466. Naturally, they resolved to have it back. Equally naturally, Bishop Watzelrode was on the side of the King of Poland (although he regarded himself as an independent ruler), and wanted to see the Order finally smashed. He devised a scheme for getting rid of the Knights by sending them to fight the Turks. The Knights regarded the Bishop as a 'devil in human shape', and prayed every day for his death; they *were* probably responsible for his death (by poisoning) in 1512. But the Bishop had his posthumous triumph; the Order was dissolved in 1525.

In the meantime, preserving the peace was a delicate business, and the Bishop had to exercise a great deal of diplomacy and persuasion, ensuring that the King of Poland was kept favourably disposed and that the Knights were kept at bay. Copernicus had to devote much of his time to writing long letters about politics. He also managed to pursue his career as a scholar, translating a somewhat platitudinous Greek manuscript into Latin, and making occasional observations of the heavens. His uncle died suddenly of food poisoning in 1512 after attending the wedding of the Polish King Sigismund, leaving Copernicus free to begin work on his first astronomical treatise,

Astronomers observing the heavens. A sixteenth-century engraving

the *Commentariolus* (*Little Commentary*), in which he first suggested that the sun is the centre of the planetary system. This failed to excite attention – not now because the human mind was unable to conceive anything so monstrous, but because the theory was offered without any kind of proof or demonstration, and no one took it seriously. Copernicus himself had little time to devote to astronomy for the next nine years, for the death of the Bishop had plunged Ermland into political chaos, and the Teutonic Knights felt this would be a good opportunity to move in again. Copernicus had finally (after fifteen years) taken up his duties as canon of Frauenburg Cathedral; as he was only one of sixteen, it is doubtful that he had been missed. His duties included administration of Cathedral lands, tax-collecting and sitting as a magistrate. This work suited Copernicus's painstaking and pedantic temperament, and he soon became the general administrator. When war broke out again between the Teutonic Knights and the Polish king, Copernicus seems to have confined himself to the city of Frauenburg (which had strong walls). He was administrator of the castle of Allenstein in 1520, but there was no question of defending it under siege; he only took part in some abortive peace negotiations.

Peace came in 1521, and Copernicus was finally able to return to his favourite hobby, astronomy. Three years later, he circulated his second work on astronomy in manuscript form – as the *Little Commentary* had also been. This was a *Letter Against Werner*, Werner being an astronomer who was still attempting to iron out the inconsistencies of the Ptolemaic system. Once again, Copernicus stated his own basic convictions: that the sun is the centre of the planetary system, that the size of the universe (which he called 'the height of the firmament') is unimaginably vast, and that the motions of the stars and planets are really due to the motion of the earth. And yet, absurdly enough, he retained Ptolemy's 'epicycles'. Why? Because he remained convinced that the path of the planets around the sun is circular. Kepler was later to realize that the actual shape is an ellipse. And it is clear that if you are expecting a circle when the planet's orbit is an ellipse, you are going to wonder why the planet seems to dawdle when it ought to be hurrying. Copernicus could see no other explanation for this than Ptolemy's 'little circles', which would cause the speed of the planets to vary. In effect, he had also adopted the gramophone record with an off-centre hole theory.

Why did Copernicus not go to the trouble of printing his ideas? The reason seems to be: timidity and caution. He had no need to be afraid of religious persecution; he lived in the age of Erasmus, in a climate of intellectual tolerance. Besides, the Pope's right-hand man, the Cardinal of Capua, Nicolas Schoenberg, had asked Copernicus for a fair copy of his book, and suggested that he ought to publish it. What bothered Copernicus was the prospect of becoming the centre of a controversy. He liked peace and quiet. It was not fear of celebrity so much as fear of ridicule. He had probably tried out his ideas on various acquaintances, and found them unreceptive. They must certainly have pointed out that if the moon is a planet, and it obviously circles the earth, then there seems to be no logical reason why the other planets should not do so. And it must be admitted that this objection looks very cogent – provided you grant the premise that the moon is a planet. It was not until Galileo pointed his telescope at Jupiter in 1610, and saw 'three little stars' in close proximity to it, that the apparent contradiction was resolved – for a planet is not at all the same thing as a satellite. Copernicus was quite satisfied to be known by repute among a few learned men.

Then, in 1539, when Copernicus was sixty-six, a young and enthusiastic scholar named Rheticus arrived in Frauenburg. His real name was Georg von Lauchen

(Rheticus means 'from the Tyrol') and he was a professor of mathematics and astronomy at Wittenberg. He was twenty-five years old, homosexual and had immense charm.

Rheticus came for a few weeks and stayed for two years. He was a Lutheran, and in Ermland Protestants were forbidden to reside on pain of death; but he was also a scholar, and so seems to have been ignored – another proof of the tolerance of the age. Rheticus set out to convince the bookish old canon that he ought to publish his theory, now written down in manuscript; his arguments were seconded by Copernicus's only close friend, Giese – once a canon, now a bishop. Copernicus tried hard to resist. He managed to persuade Rheticus to write his own *First Account* of the new system – without mentioning the name of its inventor. Having done that, Rheticus returned to the attack. He copied out the *On the Revolutions* by hand. His *First Account (Narratio Primo)* was published in Danzig in 1540, and caused an immediate stir. Now many scholars pressed Copernicus to publish. Finally, he gave way. The manuscript was handed to a printer in Nuremberg in May 1542.

Towards the end of 1542, Copernicus's health began to give way; a brain haemorrhage confined him to his bed for the remaining months of his life. He was lonely, and he was something of a valetudinarian. The loneliness was partly his own fault; the new bishop, Dantiscus, had made friendly overtures, which Copernicus had turned aside, as if afraid of the moral effort involved in friendship. After this rebuff, the Bishop pressed Copernicus to get rid of his housekeeper, a distant relative named Anna Schillings – the implication being that she was his mistress. As the Counter-Reformation gained momentum, the Church was anxious to clean up its own backyard – shutting the stable door after the horse had gone. So at the age of sixty-six, Copernicus had been obliged to dismiss his housekeeper. Life must have seemed very sad and dull during his last six months.

By this time, Rheticus had also ceased to feel friendly towards the master; the reason, fairly certainly, is that in his preface to *On the Revolutions* Copernicus inexplicably failed to acknowledge the labours of Rheticus, although he mentions Cardinal Schoenberg (the Pope's adviser) and Giese. Since Rheticus had written the *First Account* – which had spread Copernicus's ideas all over Europe – and had gone to such trouble over *On the Revolutions*, the oversight is difficult to explain. It could have been that Copernicus had turned against his disciple; Melanchthon, the Lutheran scholar, hints in a letter that Rheticus was forced to leave Wittenberg because of some homosexual scandal. Or it could have been that Copernicus was afraid to acknowledge a Protestant in a book which was dedicated to Pope Paul III; this also seems unlikely, since Rheticus was already famous as the author of the *First Account*. The likeliest explanation is that the 'oversight' was just another manifestation of Copernicus's natural lack of generosity – on a par with his failure to mention Aristarchus in the *On the Revolutions* (although his name is mentioned – and crossed out – in the original manuscript). At all events, Rheticus suddenly lost interest in the Copernican system, and when he left Nuremberg to take up a new chair at Leipzig, he handed the task of seeing *On the Revolutions* through the press to a friend called Andreas Osiander. He did not even bother to publish a biography of Copernicus that he had written in Frauenburg – clear evidence that he was embittered and disillusioned.

Opposite: Seventeenth-century celestial maps showing different views of the universe.
(Above) A Ptolemaic planisphere; (below) The Copernican world system

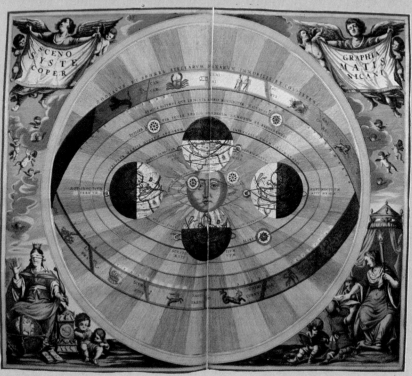

Copernicus paid for his ingratitude on his deathbed. On 24 May 1543, the first copy of *On the Revolutions* arrived. It contained an unauthorized preface by Osiander, explaining that what followed was pure hypothesis and need not be taken too seriously. What he was saying, in effect, was that the Copernican system was a mathematical theory that fitted the facts, but which need not be regarded as true. He went on to comment that the book was full of absurdities – such as the orbit ascribed to Venus, which would make that planet appear sixteen times larger when it was closest to the earth. . . . To add insult to injury, the preface was unsigned, so it seemed to be written by Copernicus himself, as if he were retracting what he had said in the book.

Legend has it that when Copernicus read the preface, he collapsed and died; the story is plausible in that he died a few hours after receiving the first copy. What seems clear is that if he had given due acknowledgement to his disciple in his own introduction (which was omitted), Osiander's preface would never have been included; Rheticus would have made certain that he checked the final printing, even from Leipzig. But it made no real difference. The canon's life's work was done, and Rheticus had played his part in sending the seed into the world. In spite of his secretiveness, pettiness and cowardice, Copernicus had lit the fuse for one of the greatest revolutions in history.

Just thirty years after Copernicus took to his bed with a brain haemorrhage, a young Dane named Tycho Brahe was crossing the grounds of Heridsvag Abbey, on his way to supper, when he stopped to stare in amazement. There was a new star in the sky – a star of exceptional brightness. Tycho was so startled that he doubted his own eyes; he called out to some peasants who were driving by to ask if they could see it too. They could. Tycho hurried off to tell his uncle, Steen Bille – the custodian of the abbey – about his extraordinary observation.

What was so astonishing was that 'new stars' were not supposed to exist; the stars were fixed and eternal. The astronomers of Tycho's time knew nothing about novae, or exploding stars. In fact, about twenty-five novae appear every year in our own galaxy, the Milky Way. But in the days before the invention of the telescope, no one noticed them. Tycho noticed the new star in Cassiopeia because it was more brilliant than a nova; this was a supernova, whose radiance was so great that it was visible even by daylight.

Other European astronomers noticed the new star; many of them insisted that it must be a comet – moving so slowly that it appeared to be 'fixed'. But Tycho was ready for this challenge. Ever since he had observed a partial eclipse of the sun, at the age of fourteen, he had been fascinated by astronomy. As the son of a wealthy nobleman – the Governor of Helsingborg Castle, facing Hamlet's Elsinore – he had the money to pay for scientific instruments. At the time of the appearance of the supernova – on 11 November 1572 – he had just finished constructing an elaborate astrolabe, with arms two metres long, and a scale graduated to minutes. One of the celebrated astronomers in Europe at the time, Maestlin of Tübingen, decided that the supernova *was* a star by holding up a piece of string so it seemed to join the new star and two on either side of it; then he tried again a few hours later, and concluded that it had not moved. Tycho went to greater lengths to decide whether it was a star or a tailless comet; he kept it under observation from its appearance in November until its disappearance in March 1574. Not only had it not moved out of line with the other stars; it showed no parallax either. And this was of considerable importance. Parallax is a body's apparent change

Tycho Brahe, the first great observational astronomer

in position when the observer changes his point of observation – as when you hold up a finger, and look at it first with one eye, then the other. As the supernova showed no parallax, it followed that it could not be fairly close to the earth – like the planets – but was about the same distance as the other stars of Cassiopeia.

In the year following the disappearance of the new star, Tycho wrote a book about his observations. It was called *De Nova Stella,* and its tables of observations instantly established his fame as one of the foremost astronomers in Europe.

Since Tycho was writing three decades after the publication of Copernicus's *On the Revolutions*, we might feel justified in assuming that he accepted the new theory of the solar system. This was not so. Oddly enough, Copernicus's book had been virtually ignored. To begin with, it was too abstract – full of tables and computations. Second, it was obscurely written. And third, the world was no more ready to accept a heliocentric theory than it had been in the time of Aristarchus. Copernicus was widely regarded as an eccentric who had produced this preposterous idea for slightly discreditable motives – perhaps a desire for publicity. So those who knew about him dismissed him as a madman.

To do Tycho justice, his prejudice against Copernicus's theory was by no means personal. It was based upon the objection that if the earth really revolved around the sun, then it would change its position by millions of miles during the course of the year. In that case, closer stars ought to show parallax against the background of more distant stars; and his most careful observation showed that they did not. Tycho had no

Tycho's world system, based on the Ptolemaic theory.

BRAHEVM, Structura EX HYPOTHESI BRAHEI IN DELINEATA.

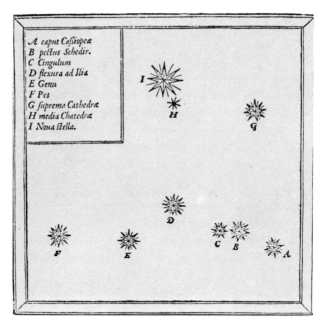

A caput Cassiopeæ
B pectus Schedir.
C Cingulum
D flexura ad Ilia
E Genu
F Pes
G suprema Cathedræ
H media Chatedræ
I Noua stella.

Tycho's drawing of the supernova of 1572. The appearance of this 'new' star challenged the assumption that the stars were unchanging. From De Stella Nova, *1573*

way of guessing that the nearest stars are so many light years away that his 'astrolabe' would have had to be two hundred times as accurate. So, on the grounds of scientific common sense, Tycho decided that Copernicus was mistaken; the earth was the centre of the universe.

It sounds as if Tycho took another retrogressive step in the history of astronomy – like Ptolemy. In fact he was, in a sense, as important as Copernicus. Tycho was one of the greatest *observers* in the history of astronomy, and his work revolutionized the science. We have already noted that the ancient Greeks were brilliant speculators but poor observers. The same is true even of Copernicus, whose famous book is based on a surprisingly small number of actual observations. It was Tycho who changed all that, and taught his fellow scientists the value of sheer hard work.

When he saw the star that made him famous, Tycho was twenty-six years old, and his life had been remarkably eventful. To begin with, he was kidnapped at the age of one. His uncle Jörgen had been childless, and had somehow persuaded his brother Otto to allow him to adopt one of his children. Tycho was born a twin in 1546; but his twin-brother was stillborn, so Otto went back on his promise. When another child was born to Otto, Jörgen abducted Tycho. Otto seems to have accepted the *fait accompli*, and his wife went on to provide him with another four sons and five daughters.

So Tycho was brought up by his wealthy uncle, and was probably spoiled – for the rest of his life he remained headstrong and addicted to his own way. At the age of thirteen, in 1559, he was sent to study at the University of Copenhagen. And in the August of the following year, he saw the eclipse that determined the future course of his life. What impressed him, apparently, was that the eclipse had been *predicted*; it struck him as 'something divine that men could know the motions of the stars so accurately'.[1] He bought himself an ephemeris – an astronomical almanac or star table –

and a copy of Ptolemy's collected works, which he proceeded to read cover to cover. He had been thoroughly bored by his existence as a son of the nobility – he spoke scathingly of 'horses, dogs and luxury' – and astronomy clearly satisfied some inborn craving. This craving was of a very rare order indeed; it was not for the romance of the heavens or the majesty of God's creation, but for precise observation. Tycho's family were statesmen and administrators, and he seems to have inherited their practicality in an unusual form.

His uncle was displeased with this unrewarding obsession with planetary tables, and sent his nephew to Leipzig University in the care of a tutor, who had instructions to try to wean him away from astronomy. Tycho had to study his star map under the bedclothes at night; understandably, his obsession only increased. His tutor did the sensible thing, and gave way. Uncle Jörgen, who was now a vice-admiral, seems to have withdrawn his objections.

It was just as well; Tycho knew already, at the age of sixteen, that he had found his vocation; as J. L. E. Dreyer puts it: 'His eyes were opened to the great fact, which seems to us so simple to grasp, but which escaped the attention of all European astronomers before him, that only through a steadily pursued course of observations would it be possible to obtain a better insight into the motions of the planets.'[2]

Yet Dreyer is here conveying – wilfully or otherwise – a slightly false impression. Tycho was not obsessed with observation simply because he happened to have that kind of temperament. He studied the heavens because, like most of his contemporaries, he was a whole-hearted believer in the doctrines of astrology. And since both Tycho and his younger contemporary Kepler were enthusiastic astrologers, it is necessary to explain briefly what it was they believed.

Astrology, as we have seen, was a legacy of the days when all religion was dominated by the earth mother and the moon goddess. As astronomy became a science, in the land of the Sumerians and Babylonians, priests observed that the path of the sun and moon through the sky – the ecliptic – is marked by twelve more-or-less regular milestones – distinct groups of stars. By the time of the Greeks, these had been given names now familiar to us: the Ram, the Bull, the Twins, and so on. Yet these constellations play no active part in astrology. They could be thought of as the figures around the edge of a clock, there for purely indicative purposes. (We could, after all, tell the time just as easily if a clock had no figures.) The sun, in its yearly round, is the hour hand. But what really counts in astrology is the planets themselves – dotted around the clock-face, so to speak. It is these, and their relation to one another, that are supposed to influence human life and destiny. Astrology is basically about the planets and their positions in the 'houses' (the twelve divisions of the clock face).

According to the astrologers, each planet has a specific type of influence. The sun stands for creativity and self-integration, the moon for instinct and feeling, Mercury for communication, Venus for sympathy and pleasure, Mars for will-power and self-assertion, Jupiter for enthusiasm and expansion, and Saturn for discipline and responsibility. The 'houses' are also associated with different aspects of a person's life: the first, with personality; the second, with possessions and money; the third, with environments and brothers; the fourth, with home; the fifth, with self-expression and children; the sixth, with servants and service; the seventh, with partnership; the eighth, with legacies and death; the ninth, with philosophy and religion; the tenth, with profession; the eleventh, with friends; and the twelfth, with limitations and enemies. So if, at birth, a man had Mars and Jupiter in his second house, it might be a reasonable assumption that he would be both lucky and determined in the art of making money.

Finally, the relationships of the planets are of central importance. An opposition – when planets are on opposite sides of the sun – indicates conflict or difficulty; a conjunction suggests harmony. The various other aspects – squares, trines, sextiles and so on (respectively 90°, 120° and 60°) indicate various degrees of conflict or harmony.

These explanations are necessary to explain why Tycho was so excited by a celestial event which took place when he was sixteen; in August 1563, a conjunction of Saturn and Jupiter was due to take place – which is to say that Jupiter was due to eclipse Saturn. The conjunction of two such planets – of opposite qualities – must have suggested interesting possibilities. On the personal level, for example, it could have meant that enthusiasm and expansion were about to be curtailed by the hand of discipline. And this, as a matter of fact, was precisely what was about to happen to Tycho; Uncle Jörgen was about to send him to Leipzig with a tutor entrusted with the task of persuading him to forget about the planets. We do not know what significance Tycho attached to the eclipse, but we know that, between 17 and 24 August 1563, he observed the planets minutely, measured their angular distance with a home-made instrument, and discovered that both of his sets of planetary tables were in error. And if he read the signs as forewarning him of his uncle's growing resistance, he must also have comforted himself with the thought that it was Jupiter – the planet of enthusiasm – that was occluding Saturn, and that therefore a certain determination and strength of purpose should see him through. As we know, his tutor, Anders Vedel, finally gave up the attempt to make him forget astronomy, and the two became close friends.

In June 1565, less than a month after Tycho's return from Leipzig, Jörgen died of pneumonia – he had dived into a river to save King Frederick II. Thereupon, Tycho went back to his travels, and spent the next nine years studying at various universities – Wittenberg, Rostock, Basle and Augsburg. At Rostock, he had the misfortune to lose the bridge of his nose in a duel, and had a replacement made of gold and silver. He seems to have taken the loss in good part, and became friends with his adversary.

Tycho's father died when he was twenty-four; in the same year, Frederick II promised him some kind of sinecure – perhaps a post in the Church, like the one that had been given to Copernicus. Tycho and his brother Steen inherited their father's fortune, and he moved to Heridsvag Abbey to live with his uncle Steen Bille. It was here, as we have seen, that he observed the 'new star', wrote his book about it, and achieved celebrity among European scholars.

Two years after the publication of his book, Tycho made another tour of Europe, and stayed with the Landgrave William IV in Cassel. William was an enthusiastic astronomer who had observed the new star from his own observatory. When Tycho returned to Denmark, the Landgrave sent a message to the King, urging him to provide the funds for Tycho to build his own observatory. Frederick offered Tycho the use of various castles, which he declined, for he had fallen in love with the city of Basle and intended to move there. Frederick decided that a man of such brilliance should stay in Denmark, and asked Tycho to come and see him. On 11 February 1575, the King offered Tycho a beautiful island called Hven, in the sound between Elsinore and Landskrona, together with a pension and various sinecures. He was also offered the island's rents for his own use. Tycho accepted. And the world's greatest astronomer proceeded to build the world's first modern observatory, the Uraniborg. This was a

Venetian zodiac, by G. B. Agnese. Astrology continued to flourish in the Renaissance

fantastic castle in the middle of magnificent grounds; and for the next twenty years Tycho ruled his island domain, a mixture of ascetic scholar and oriental despot.

The Uraniborg sounds like Kubla Khan's stately pleasure-dome. The rooms had running water, and the sixteenth-century equivalent of telephones – statues with speaking tubes. Tycho took a childish pleasure in the communications system. He might remark casually that it would be pleasant to have some particular dish, and would surreptitiously pull a hidden lever; a moment later, a servant would enter carrying the dish. The megalomaniac genius printed his own atrocious poetry on his own printing press. He gave huge banquets, during which he ate and drank like a character out of Rabelais. He took a peasant woman to his bed, and insisted that she be

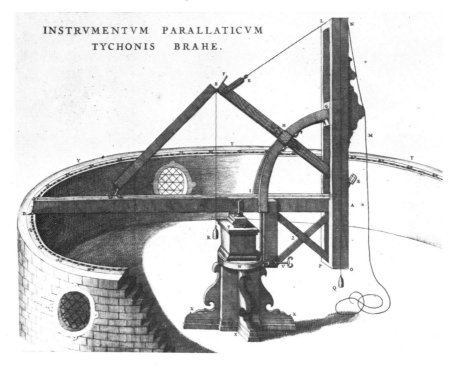

Instrument built by Tycho Brahe which he used to draw up his star catalogue

treated as the mistress of the house. His main study was decorated with a mural showing the eight great astronomers of history – culminating, as was only proper, in Tycho himself.

Despite all the banqueting and entertainment, Tycho worked endlessly at his observations. They were incredibly accurate, in spite of the lack of all optical instruments. He measured the length of the year with an error of only one second. He created new star tables, and made adjustments for the refraction of star light in the earth's atmosphere. His observations of the planets were the most accurate ever made. Arthur Koestler conveys something of his achievement when he says (in *The Sleepwalkers*) that while the observations of previous astronomers were like still photographs, Tycho's were like a cinematograph film: that is, there were so many that they virtually formed a moving record. When a comet appeared in the sky in 1577, Tycho observed it closely, and wrote a book in which he proved that it could not be within the earth's atmosphere, because the difference in parallax between Hven and Prague showed it to be much further away than the moon. It was this kind of simple observation – comprehensible to anybody, yet contrary to all the accepted notions of his time – that made Tycho famous all over Europe.

Unfortunately, Tycho's flamboyant style slowly developed into delusions of grandeur. He treated his tenant farmers appallingly, demanding more than his due, and putting them in prison if they objected. He was overbearing and rude – even to the young king, Christian IV, who replaced his father in 1588. Moved by the protests of the farmers, the King reduced Tycho's income. And Tycho, who had always been torn between his love of Uraniborg and his love of travel, decided to move on. At the age of fifty, in the Easter of 1597, he left Hven with an immense entourage, and cartloads of instruments and books – he even took the printing press. From Rostock, in Germany, he wrote Christian an impertinent letter, complaining of his treatment, and offering to return if he made amends. The King, with singular reasonableness, wrote back a letter in which he refuted Tycho's accusations in detail, and added that Tycho was welcome to return if he showed more respect. Tycho, worried by the plague that had arrived in Rostock, moved on eastwards.

He spent the next year at a castle in Wandsbeck, near Hamburg, placed at his disposal by a rich admirer, then spent the following winter in Wittenberg. He was headed in the direction of Prague, where he hoped to obtain the patronage of another monarch – Rudolph II, son of Maximilian II, Archduke of Austria, King of Bohemia

Uraniborg, the 'Castle of the Heavens', Tycho's great observatory at Hven

May signals the triumph of Apollo over Venus. Allegorical fresco, Ferrara

and Hungary, and Holy Roman Emperor. By way of introduction, he sent Rudolph a copy of his star catalogue, specially dedicated to the Emperor. In June 1599, the Emperor and the astronomer finally met, and Tycho was given a pension of 3000 florins a year, and a post equivalent to Astronomer Royal (it was called Imperial Mathematicus). He was also given the castle of Benatsky, on a hill overlooking the River Iser.

Tycho's important work in astronomy was finished, and he had only two more years to live. But he still had one important service to perform for science: to launch a young mathematician named Kepler on a career that would make him the greatest astronomer in Europe.

Johannes Kepler, author of the Laws of Planetary Motion

Johannes Kepler was twenty-nine when he met Tycho Brahe in 1600, and he already regarded himself as an accomplished astronomer. Yet the book on which he believed his fame would rest was an exposition of a crank theory of the solar system. If Kepler had died before he met Tycho, his name would now be unknown.

He had been born in 1571 in the little town of Weil in Swabia, near the Black Forest, a skinny, sickly, ugly child, whose mother was a witch and whose father was an adventurer with a criminal streak. His grandfather had been mayor of Weil, but since then the family fortunes had gone downhill – perhaps because of the strain inherited from his grandmother, whom Kepler described as a liar, an inveterate troublemaker and a bearer of grudges. During his childhood, his father was away much of the time, fighting as a mercenary; at one point he narrowly escaped hanging for some un-specified crime. When Kepler was seventeen, he deserted his family and was never heard of again.

Kepler was ill throughout much of his childhood; he was also accident-prone. But at least he was lucky in the place of his birth; Swabia had an excellent educational system, with scholarships for the children of the poor. Kepler's days at elementary school were academically successful, if not particularly happy. In fact, he was loathed by most of his schoolfellows; he was physically unpleasing (admitting to a 'dog-like horror of baths'), a sneak, a toady, an opportunist and a bookworm. He combined a passionate desire to be liked with a conviction that he was thoroughly unlikeable. His school-fellows concurred, and frequently beat him up. His only compensation lay in his cleverness, and even that was of the irritating kind that enjoys showing off. Altogether, it was a purgatory of a childhood.

Yet there was a toughness and determination about Kepler that made his misery fruitful. It made him self-analytical, and sharpened his intellectual powers. It also meant that he turned to the world of ideas for escape. In the nineteenth century he would have been a typical romantic, wandering in the countryside with a book in his pocket. But in the sixteenth century books were rare and expensive, and nobody had yet noticed that nature was beautiful and romantic – that would not happen for another two centuries – so Kepler had to find solace in mathematics and astronomy. His mother had shown him the comet of 1577 (about which Tycho had written his book) and an eclipse of the moon in 1580. But Kepler lacked Tycho's passion for observation; his talent was for theorizing. He also wrote verses and tried his hand at comedies. He read Aristotle in Greek, but seems to have failed to develop a taste for philosophy; he irritated his fellow students by telling them that the study of philo-sophy was a symptom of German decadence.

The aim of the dukes of Württemberg in creating the excellent educational system was to recruit clever young men for the Protestant clergy. Kepler was a natural choice, and at thirteen he was sent to a theological seminary. It was while he was there that he encountered the theory of Copernicus; predictably – for he was nothing if not an intellectual rebel – he was at once converted, and defended Copernicus in a public disputation. Yet astronomy remained one interest among many; he had no thought of making it a career.

At the age of seventeen he went to Tübingen University; he graduated at the age of twenty, then went on to the theological faculty, where he studied for another four years with a view to entering the ministry. Then fate intervened. In Graz, capital of the Austrian province of Styria, the 'mathematicus' of the Protestant school died, and the governors asked Tübingen to recommend a replacement. The university authorities suggested Kepler – possibly because someone had already begun to wonder whether he would make a good clergyman.

At first Kepler was inclined to refuse. At twenty-three, he had no intention of devoting his life to mathematics and astronomy (these being the two chief subjects required of a mathematicus). Also, he probably liked the idea of being in charge of his own little flock. But then there was no reason why he should not go back to theology in due course, and he was ready to go out into the world. He still lacked any sense of definite direction, and was willing to wait and see what fate had to offer. So in April 1594, he arrived in Graz and became a schoolmaster. But he was soon bored and dissatisfied. His class was small, and in the second year there were no enrolments at all. Graz struck him as unbearably provincial. Kepler always found it difficult to get along with people, and imagined himself rather more disliked than he actually was. (The school records show that the governors actually regarded him with approval.)

During his first year, Kepler inadvertently made himself a reputation as a prophet. Part of his duty as mathematicus was to publish a kind of *Old Moore's Almanac*, explaining what the stars foretold, and he prophesied a spell of intense cold and a Turkish invasion. Both happened on cue. Yet Kepler was not an enthusiastic astrologer; he felt – as any modern student of the subject feels – that the whole 'prophetic' side of astrology was nonsense. He was firmly convinced that the heavens influence the earth, writing: 'Nothing exists nor happens in the visible sky that is not sensed in some manner by the faculties of Earth and Nature. . . .'[3] He also wrote: 'That the sky does something to man is obvious enough; but what it does specifically remains hidden.' In short, he believed that the planets influence human temperament, but not necessarily that they can cause specific events. The kind of astrology he was being asked to do – and continued to do for the rest of his life – was the equivalent of 'What the Stars Foretell' in a modern newspaper. Yet this belief in the influence of the planets led Kepler to scrutinize the heavens with the feeling that somehow they held the key to the mystery of human existence.

On 9 July 1595, just over a year after his arrival in Graz, Kepler was drawing on the blackboard a geometrical figure: it was a triangle enclosed in a circle, with another circle inside the triangle. Suddenly, he was struck by a revelation – or, at least, what he thought was a revelation. He had been brooding for some months on the distances of the planets from the sun; why were they not spaced neatly and symmetrically, as one might expect? Today we know the answer: that it depends on the speed of the planets and the laws of gravity. But Kepler knew nothing of the laws of gravity. Now he saw what he thought might be a solution. If an equilateral triangle (one with three equal sides) is inscribed in a circle, and another circle drawn inside, the relationship between the size of the two circles is always the same, no matter how big they are. And this relationship, Kepler suddenly realized, is the relationship between the orbits of Saturn and Jupiter.

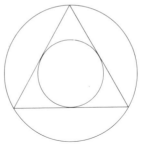

Could that, he wondered, be the answer to the puzzle of the orbits – that God so designed them that neat geometrical figures can be fitted inside them? Not, of course, a series of triangles inside circles – that would be too unimaginative (and besides it didn't fit). But squares, octagons, and so on? Or, better still, cubes, octahedrons and so on – for after all, space is three-dimensional. . . . Convinced he had received a revelation, Kepler proceeded to work it out. What three-dimensional solids can be inscribed inside a sphere so that the vertices touch the shell of the sphere? There is the tetrahedron (a pyramid made up of four equilateral triangles), the cube, the octahedron, the dodecahedron (with twelve sides) and the icosahedron (with twenty sides). And when the calculation was finished, Kepler was convinced that he had found the answer to the design of the solar system. The cube fits between the orbits of Saturn and Jupiter, the pyramid between Jupiter and Mars, the dodecahedron between Mars and earth, the icosahedron between earth and Venus, and the octahedron between Venus and Mercury. It fitted! The five 'perfect solids' of Pythagoras fitted the five orbits between the outermost and innermost planet. (At this time, of course, only six

Ceremonial Hall of the Hradschin Palace, Prague, where Kepler worked as Imperial Mathematicus

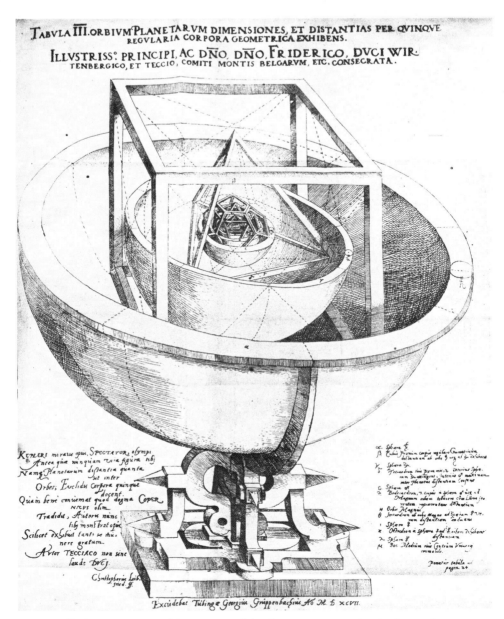

Kepler's explanation of the structure of the planetary system using the five regular polyhedra between the spheres of the planets. From Harmonice Mundi, *1619*

planets were known – Uranus, Neptune and Pluto were still to be discovered.) In great excitement, Kepler produced a book, the *Mysterium Cosmographicum*, to expound his theory. It appeared when he was twenty-five, and spread his name among European scholars. (He took care that everyone of note should receive a copy.) It was printed by his Alma Mater, Tübingen, under the supervison of his old astronomy professor, Maestlin (the same who had measured the new star and its neighbours with a piece of string).

When Kepler received his first copies, he had just married a miller's daughter named Barbara Muehleck; she was a fat, melancholy woman who had been twice widowed. Kepler cast a horoscope of their union, and decided that it was thoroughly unpropitious; but it was too late to back out. The marriage was even worse than he expected. She was mean, stupid and bad tempered – although, apparently, a doting mother. She already had a number of sickly children by an earlier marriage, and she bore Kepler five more, three of whom died. She became increasingly distraught and depressive, and died after fourteen years of marriage in a state verging on insanity.

Most of Kepler's biographers have made the obvious comment that he revived the Pythagorean approach to astronomy, complete with mathematical mysticism. But this is not entirely accurate. Pythagoras, in the excitement of geometrical discovery, had leapt to the conclusion that the secret of the universe lay in whole numbers; it was a form of wishful thinking. Kepler, although he had an element of crankiness in his composition, was a practical astronomer. He *knew* there had to be a reason for the planets to be such unsymmetrical distances from the sun, and he was right. But he made the false – if reasonable – assumption that their orbits are perfect circles. Staring at a diagram containing these circles, he could see that there had to be some hidden relation between their diameters. But what could it be? His notion about the 'perfect solids' was, at worst, an inspired guess. If we recall that every astronomer since Plato had taken it for granted that the heavens were made up of a series of invisible spheres, it was by no means a preposterous idea.

In the *Mysterium*, Kepler admitted that there were a few minor discrepancies between his theory and the actual figures. Yet he remained convinced that he had stumbled on the great secret of the universe, and that a little further calculation – and observation – would reveal the fundamental truth of his inspiration. After publication of the book, he settled down to the serious study of mathematics, convinced that this would finally remove the discrepancies; but he was disappointed. No matter how he juggled the figures, the distances and velocities of the planets refused to fit into his symmetrical moulds. It was clear that the theory lacked the necessary accuracy to account for the functioning of the universal clockwork. What he needed was more data, more actual observations. And the only man in Europe who could supply those was on the island of Hven, in Denmark.

In fact, Tycho Brahe had received Kepler's *Mysterium* with kindly approval, and invited the young man to come and stay with him some time; but six hundred miles had been too far to travel. Tycho, however, was not a man freely to offer his own observations to rivals; where his 'discoveries' were concerned, he was something of a miser. Kepler told Maestlin that one of these days he would have to try to 'wrest [Tycho's] riches from him'.[4] There seemed then to be no immediate prospect of a meeting.

Then, in the summer of 1598 – by which time Tycho was already on his way across Europe – fate took a hand, and shook Kepler out of his academic security. Unlike his native province of Swabia, Austria was still a Roman Catholic country; Protestantism was tolerated by its rulers – the Hapsburgs – but not much liked. The young Archduke Ferdinand of Styria suddenly decided it was time to take a stand, and ordered the Protestant school at Graz to be closed; a few months later he ordered all Protestant teachers to leave the country on pain of death.

Oddly enough, Kepler found that he was the only teacher exempted from the order. He was something of a celebrity; his subject – mathematics – was regarded as theologically neutral; and he had friends in high places (the chancellor of Bavaria was an admirer). All the same, with the school closed down and an edict against Protestants, it was obvious that he had no future in a Roman Catholic principality. The death of his daughter from meningitis – only a year after the death of her brother – deepened his depression, particularly as he was fined for burying her with Protestant rites. Kepler decided to move on. And once more fate intervened; a councillor of the Emperor Rudolph happened to be in Graz, and offered to take Kepler in his suite back to Prague – where Tycho had just been appointed Imperial Mathematicus. Kepler set out for Prague on 1 January 1600, and the new century began in more ways than one.

Apart from the fact that both were men of genius, Tycho and Kepler had little enough in common – except shortness of temper. Tycho was wealthy, famous, and used to carrying all before him. Kepler was poor, harassed, and accustomed to bad luck. Moreover he had a touch of paranoia that was bound to be aggravated by Tycho's overbearing manner. In fact, the two men seemed made to detest one another. Yet it was an important factor that had drawn them together: they needed one another. Kepler's speculative genius was being forced to mark time for lack of necessary data. Tycho had the data; but he was no mathematician; also, he was urgently in need of an assistant, some of his former assistants having deserted him. Kepler was handed the task of studying the planet Mars.

His early days at Benatsky Castle were spent in a state of intellectual shock. For the past six years he had worked in isolation – a loner, working with inadequate data. Suddenly, he had data by the cartload. And it led to problems that were so great 'that I nearly went out of my mind'. But he saw at once that Tycho lacked the organizing intellect to make proper use of all this material, and that he, Kepler, was destined to be the architect of the grand design.

Tycho himself realized this, and it made his attitude to Kepler highly ambivalent. He clung to his data like a miser, releasing it in dribs and drabs, sometimes in a casual remark made at meal times. Kepler was desperate but determined; his mind locked onto the problem, and bulldog-like he tugged grimly. Tycho was not the only one who was jealous of Kepler; his family and assistants disliked the upstart. Fortunately, Kepler's control of his temper was so poor that Tycho was placed in the unlikely position of the magnanimous and long-suffering father figure. After two months, Kepler exploded, called Tycho unforgivable names, and stormed off to Prague to stay with the councillor who had brought him there. Then he became a prey to feelings of guilt, so that when Tycho went to fetch him, he returned with his tail between his legs. The metaphor is appropriate; Kepler still liked to compare himself to a dog.

Kepler's family had remained in Graz; shortly after the reconciliation, he decided to return to fetch them. Secretly, he was hoping that the sectarian squabbles had blown over, and that he might stay on there. In fact, things were worse than ever. The Protestants of Graz had all been given an ultimatum: return to the fold, or get out. It applied even to Kepler. There was nothing for it but to return to Prague – this time with wife and children, but without his household effects, as he had no money for their transport.

Kepler was ill with fever, but Tycho pressed him to begin work immediately on a pamphlet attacking an astronomer known as Ursus. Kepler loathed the drudgery and, when his father-in-law died six months later, took the opportunity to go back to

Tycho Brahe in his observatory at Hven

Graz for his wife's share of the inheritance. Nothing much came of this, but his former fellow citizens had heard that he was now in the employ of the Holy Roman Emperor and he spent several months being fêted and spoiled. He was back in Prague in August, his health miraculously restored by all the attention.

Two months later, in October 1601, Tycho dined well at a nobleman's table, and made the mistake of waiting until he got home to urinate. Infection produced a fever and after five days of pain he died 'among the consolations, prayers and tears of his people'.[5] He was fifty-four years old. Kepler was no doubt among the people who shed tears; his gloom must have been augmented by the thought that he might soon be homeless. Meanwhile, the Emperor, that vague and unworldly man, wondered whether the money he had invested in Tycho – and in the star charts he was constructing – would be wasted. From a social point of view, Kepler was no Tycho; but he was obviously the only man qualified to replace the Imperial Mathematicus. So, two days after Tycho's death, Kepler's anxieties were laid at rest, and he was appointed in his place.

It was unfortunate that Kepler was the sort of person for whom every bright cloud has a coal-black lining. In this case, the drawback was the Emperor himself. Rudolph II might have been perfectly happy as a Renaissance prince living in a small domain; he was completely unsuited to governing an empire. He was a scholar and a recluse; he was subject to fits of depression, and his health was poor. He was an ardent student of alchemy and occultism, and had played host to the famous Dr John Dee and his dubious famulus Edward Kelley. But whatever his qualifications as a scholar, he was not the man to hold together an empire split by religious dissension. By the time he appointed Kepler to Tycho's place, Rudolph felt himself hemmed in by enemies and critics. He knew that the Hapsburg archdukes would have preferred to see his brother Matthias on the throne; and eventually he would hand over Austria, Hungary and Moravia to Matthias. By the time he died, in 1612, he was virtually a prisoner in his own palace. Under the circumstances, it can be understood why

Rudolph II, the eccentric Holy Roman Emperor
who encouraged astronomical science

he was an unsatisfactory master, and why Kepler's salary was always in arrears.

Kepler also had trouble with another court official, a Junker named Tengnagel, who had once been Tycho's assistant, and had married Tycho's daughter. Tengnagel had seized all Tycho's instruments after his death, and denied Kepler access to them. Tengnagel also wanted to get hold of Tycho's books of observations on the planets; but Kepler had laid hands on these, and declined to part with them. It was just as well; the observations would have been no use to Tengnagel, and the theft enabled Kepler to write his classic work *Astronomia Nova de Motibus Stellae Martis ex Observationibus Tychonis Brahe – The New Astronomy or Physics of the Skies*. But meanwhile Tengnagel made trouble, and even tried to steal the papers back again. The Emperor 'bought' Tycho's instruments from Tengnagel (he never paid for them) but, typically, kept them locked away, where they gradually rusted and became useless.

It was fortunate that Kepler now had what he needed to continue his work: Tycho's observations on Mars. With these, he could draw a sketch-map of the planet's orbit. And, having done so, he found himself faced with an intellectual road-block. The observations revealed that the orbit of Mars was lop-sided. Or, more precisely, that it was egg-shaped.

Copernicus had also recognized that the planets do not move in perfect circles; he had tried to explain this by saying that the centre around which they revolve is not the sun itself, but a point in space some distance away – like a record with an off-centre hole. Now Kepler's calculations showed him that it was worse than that. The orbit of Mars was not a lop-sided circle, but an ellipse. And Kepler's mind found this impossible to accept. The epicycles of Ptolemy and Copernicus were unsymmetrical enough, but this egg-shaped orbit was preposterous.

An ellipse is a figure made by giving a 'circle' two centres instead of one. Anyone can draw a circle with a pin, a loop of cotton and a pencil, stretching the cotton between the pin and pencil. But if a second pin is stuck inside the loop, the circle now becomes an ellipse. Kepler knew about ellipses; but he could not accept that the orbit of Mars was a genuine ellipse; he thought it must be a cross between an ellipse and a circle. He could simply not believe that the solution could be so straightforward. We must bear in mind that at this stage of intellectual history, no one had yet thought of the idea of gravity, that elastic string on which the sun swings the planets around its head. Kepler realized that some invisible 'force' was involved; but he thought of it as being more like a whirlpool, with the sun at its centre. As to why the planets should move in irregular orbits, he could only make a wild guess: that they have north and south poles, like magnets, and that when one pole points at the sun, it is attracted, whereas when the other does the same thing, it is repelled. It was an intelligent guess; but the magnetism would have had to have been many times as strong to account for the orbit of Mars.

Besides, there was another problem. Tycho's figures also showed that Mars had different speeds. It went faster as it approached the sun, and slower as it travelled away. This problem baffled Kepler for the best part of a year until one day, staring at his diagram, he was struck by an insight. If we imagine the planet connected to the sun by a long piece of elastic it is obvious that the elastic will eventually sweep out the whole area of the ellipse. When it is close to the sun, it moves faster, so the area it will sweep across in, say, one day will be short and fat. At the other end of the 'egg', the area it will sweep out in a day will be long and thin (because the planet is moving slower). Was it possible that the actual *area* it could cover in both cases could be the same? A check revealed that this was so. The area covered in a definite time was always the same. Kepler had discovered the basic law of planetary motion.

Oddly enough, he had still not decided on the exact shape of the orbit; he only knew it was roughly elliptical. Again, there were months of struggle and frustration,

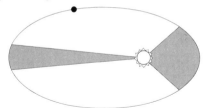

during the course of which he stumbled on an equation linking the flatness of the orbit with the distance between its two 'centres'. Now, at least, he had an equation describing the orbit of Mars. When this seemed to take him no further, he abandoned it, and tried out the hypothesis that the orbit might be a straightforward ellipse. It only dawned on him after months of work that his equation *was* the formula for an ellipse. And although this discovery came after the one about the area covered in equal times, it obviously came first in order of precedence. So it became Kepler's celebrated First Law of Planetary Motion: that the orbit of a planet is an ellipse, of which one focus is the sun. The Second Law – that a string joining the sun and planet sweeps over an equal area in equal times – followed in logical sequence. But still one major problem remained, the problem he had tried to solve in his first book – why are the planets such unsymmetrical distances from the sun?

But now he had the clue that had eluded him ten years earlier. In those days he had thought solely in terms of space, and the shapes that might fit into it. Now he knew that *time* is just as important in this equation. This time the solution took many years – it came to him in 1618, nine years after he had published his first two laws in *The New Astronomy*. Staring at the tables that showed the distance of planets from the sun, and the time it took each one to revolve around it, it came to him that the cube of the distance is equal to the square of the time it takes. (This – we may recall – was the phrase Swift was to use in writing the passage about the satellites of Mars in *Gulliver's Travels*.) It is the famous Third Law of Planetary Motion. Kepler illustrates this law (in a work called *Harmonice Mundi, The Harmony of the World*) with the example of Saturn, whose distance from the sun is about nine and a half times that of the earth, and whose year is about twenty-nine and a half earth years. Nine and a half cubed (multiplied by itself three times) gives approximately 867, which is also the figure we get if we square twenty-nine and a half.

So with Johannes Kepler the science of astronomy finally comes of age; it ceases to be philosophical speculation and guesswork. Kepler, of course, did arrive at his conclusions by speculation; but it was speculation based upon the precise measurements of Tycho Brahe. The coming together of Tycho and Kepler was more than a momentous historical meeting; it was a symbol of the method of modern science – a method that would cause the world to change more in three centuries than in the previous three million years.

It would be pleasant to record that the publication of Kepler's *The New Astronomy* brought him recognition from the four corners of the earth. In fact, like Copernicus's *On the Revolutions*, it was hardly noticed – even in the world of scholarship. The reason was much the same: it was highly technical and difficult to follow. Besides which, no

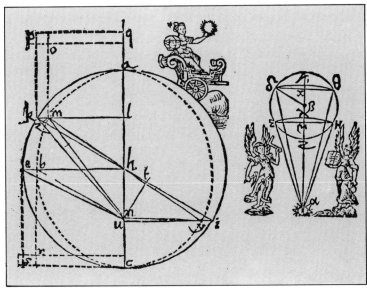

Kepler's explanation of the elliptical orbit of Mars. From Astronomia Nova, 1609

one felt that these laws of planetary motion were particularly important. What really seems to lie behind the indifference is that basic human craving for simplicity which led to the rejection of Aristarchus. Common sense tells us that the planets ought to travel in circles around the sun; so *why* should they travel in ellipses instead? That they do seems to be a kind of cussedness, an attempt to confuse us.

The reason, in fact, is simple. Two forces act on a planet: the force of gravity, trying to drag it into the sun, and its centrifugal force as it swings around in its orbit. If this latter force was too great, the planet would fly off into outer space. If it was too weak, the planet would fall into the sun. If it was *precisely* great enough to stop the planet falling into the sun, then the planet's orbit would be circular. But it is stronger than this; it tugs away, like a child tugging at its parent's hand as they walk along the street. We can see that if the parent had a rubber arm, it would stretch and then contract again. The 'rubber arm' of gravity does the same thing; the planet gains speed as it approaches the sun, then is grabbed by the force of gravity and made to do a U-turn round the sun. Now its additional speed – which it picked up approaching the sun – causes it to try to shoot off in a straight line, and the sun has gradually to slow it down. Finally, too slow to resist any more, it almost comes to a halt, and once again turns back towards the sun. . . . In addition to all this, we have to take into account the gravitational pull of the other planets, also trying to tug the body off-centre. When all these factors are taken into account, we realize that it would be incredible if any planet or satellite *did* revolve in a circle. But in order to see this, we have to think in terms of a complex web of forces, not some idealized system of geometrical shapes. Kepler began by thinking in terms of idealized shapes; and without Tycho's observations he would probably have continued to do so for the rest of his life. So perhaps it is Tycho – the man without a single important idea or discovery to his credit – rather than Kepler who should be regarded as the father of modern mathematical astronomy.

The New Astronomy had been completed in 1605, but was not published for another

117

four years – Rudolph's throne was tottery and his treasury was empty. By 1611, it was clear that the Emperor was losing his mind, and he was forced to abdicate in favour of his brother Matthias, who had little use for astronomers. Plague and civil strife came to Prague. Kepler's wife Barbara died; so did his six-year-old son, his favourite. In 1613, Kepler married Susannah Reutinger; this time he seems to have made a more satisfactory choice, and they remained happily married for the seventeen years left to him; she bore him seven children.

By the time of this second marriage, Kepler had secured for himself a job as a professor of mathematics in Linz in Upper Austria. It was dull and provincial, but fairly peaceful; and at least his salary was regularly paid. The only cloud on the horizon was the refusal of the local parson to grant him communion, on the grounds that his theology was unsound.

In 1615, domestic troubles darkened the horizon: his mother was accused of witchcraft. She seems to have been a malicious and disagreeable old lady, and there may have been some truth in the accusation – she had been brought up by an aunt who was burned at the stake. These problems dragged on for another six years, and cost Kepler an immense amount of energy and anxiety. It was largely through his efforts that she was acquitted in 1621.

In 1618, the bitter conflict between Roman Catholics and Protestants – which had already driven Kepler out of Graz – exploded into the Thirty Years' War. As Ferdinand's campaign against Lutherans and Calvinists intensified, the Protestants of Bohemia rebelled, and threw two leading Catholic ministers out of the window of Prague Castle. Suddenly, Europe was at war. In the same week as the 'defenestration of Prague', Kepler completed the book that he regarded as the crowning achievement of his life's work: *The Harmony of the World*. He wrote: 'Having perceived the first glimmer of dawn eighteen months ago, the light of day three months ago, but only a few days ago the plain sun of a most wonderful vision – nothing shall now hold me back.'[6] The 'wonderful vision' seems to have come to him in a flash, like the revelation about the five perfect solids of twenty-five years earlier. It was the notion that Pythagoras was correct after all: the universe *is* governed by the laws of music.

We may recall that Pythagoras discovered the connection between the length of a string and the note it gives; and that if one string is twice the length of another, its note will have just half as many vibrations. (And that the difference between the two notes will be precisely one octave.) Kepler had never quite given up the notion that the answer to the proportions of the universe would be found in music. Now the insight flashed upon him that God had coded this musical proportion into the velocities of the planets round the sun. For example, the maximum speed of Saturn is 135 seconds of an arc, and its minimum speed 106 seconds. The ratio of 106 to 135 is very nearly the ratio of four to five – which, according to Pythagoras, is a major third. By the same method, Jupiter's 'code' proves to be a minor third, and that of Mars, a fifth. The 'tune' played by the earth is Mi, Fa, Mi, which Kepler interpreted as Misery, Famine and Misery. . . .

Understandably, most commentators dismiss *The Harmony of the World* as the aberration of an overworked mathematical genius. But the case of Poe's *Eureka* should have alerted us to the danger of these hasty intellectual judgements. Here again, we have a work whose title-page might have borne the epigraph: 'What I propound here is true.' The simplest explanation would be that Kepler felt instinctively that he had still failed to lay hands on the secret that would resolve the universe into harmonious mathematical formulae – the secret we now know to be gravity. But this is to leave out of account the cornerstone of Kepler's belief in the relation of heaven and earth: his feeling that the planets somehow influence the affairs of men. We know, of course, that he took a poor view of the kind of astrology on which an Imperial Mathematicus was expected to waste much of his time, calling it 'the foolish step-daughter of astronomy'.[7] Yet he *was* convinced that the position of the planets at the moment of birth exerts an influence on human temperament, and therefore on human destiny. We know that he made his reputation in Graz at the age of twenty-three with a remarkably accurate prophecy of 'unheard-of cold' and a Turkish invasion. We also know that in his days in Prague, Kepler was asked to cast the horoscope of a young nobleman whose identity was withheld, and that the accuracy of his character-analysis astounded the man for whom it was intended – Albrecht von Wallenstein, the future imperial commander-in-chief. Sixteen years later, Kepler was asked to bring Wallenstein's horoscope up to date; he did so, but stopped in 1634, a year that would bring 'dreadful disorders over the land'. Wallenstein was murdered in that year.

This was the underlying certainty that made Kepler declare: 'Yes, I give myself up to holy raving. I mockingly defy all mortals with this open confession: I have robbed the golden vessels of the Egyptians to make out of them the tabernacle for my God . . .'[8] He had spent eighteen years of his life brooding over figures and planetary orbits; yet he remained as deeply convinced as ever that he was dealing with something more than clockwork: that in some sense, the universe was intelligent as well as harmonious. Kepler, the first great astronomer of the new age, felt in his bones that sense of meaning that led Neolithic man to erect Stonehenge and Carnac. It was this recognition that made the composer Hindemith turn the life of Kepler into an opera called *The Harmony of the World*, at the end of which the planets sing that mortal beings can only feel, but not understand, the overruling cosmic harmony. As a comment on a man who had spent his life trying to understand the laws of the universe, this statement has an extraordinary penetration.

Even after publication of *The Harmony of the World*, Kepler still had important work to do. He wrote a vast *Epitome of Copernican Astronomy* (1618–21), in which he extended his laws of motion from Mars to all the other known planets. This is the first work of astronomy to reveal the universe as it is seen by the modern astrophysicist. In 1627, he finally completed the work that had been commissioned by the Emperor a quarter of a century earlier, the great 'Rudolphine' star tables, whose precision gave a new impetus to astronomy during the next century. It was just in time for the annual Frankfurt book fair, where, nineteen years earlier, a man had exhibited a device that was to revolutionize astronomy: the telescope.

By the time the *Rudolphine Tables* appeared, Kepler had again been driven from his home by the religious antagonisms of his time; Linz had been besieged and partly burned down by Lutheran peasantry. Kepler spent the last three years of his life in restless wandering: from Ulm to Prague, where he became Court Astrologer to Wallenstein; to Sagan, where Wallenstein lived; and finally, after Wallenstein's dismissal by the Emperor, to Leipzig. In spite of his European fame, and many invitations to teach at universities abroad, he still suffered a deep sense of insecurity, dating from his early years. He had travelled to Ratisbon, hoping to persuade Emperor Ferdinand to pay some of the 12 000 florins he still owed him, when he fell ill with a fever. He died on 15 November 1630. In a horoscope he cast for himself for that year, he had noted that the planets were now in the same position as in the year of his birth. . . .

Galileo Galilei, physicist, brilliant astronomer and bellicose champion of the Copernican system

CHAPTER FIVE

THE LAWGIVERS

I t is a pity that one of the most important events in European history should have gone completely unrecorded. But at least we know roughly when it took place, and the name of the person concerned; the rest can be surmised. Around the year 1290, a Florentine nobleman named Salvano degli Amati seems to have been involved in the manufacture of glass – a commodity that had been known since pre-Roman times. Amati would have been aware that a crystal ball will enlarge the fingers of the hand in which it is held, and that if a drinking-glass is placed on a manuscript, the letters become distorted – and magnified. There is evidence that short-sighted monks used such devices to read manuscripts. Amati seems to have been the first deliberately to manufacture concave circles of glass for purposes of magnification. But these magnifying glasses had one disadvantage: they had to be held above the manuscript. Amati made the further discovery that they worked just as well if they were held in front of the eye instead. So he took the logical step of enclosing them in a frame made of leather, which could be attached to a hat or cap. He had invented spectacles, and his achievement was duly recorded on his gravestone, in the Church of Mary Magdalen in Florence, when he died in 1317. It is also stated that the inventor of 'occhiali' (as he called them) kept the process by which they were manufactured a secret, which suggests that he sold them.

The world was ready for this tremendous invention. In the Dark Ages, it had only been monks who had to wear out their eyesight over manuscripts. But now businesses were flourishing, every merchant had to keep books, and the new universities were turning out thousands of graduates who would devote their lives to study. An Arab physician of the time

remarks sadly that shortsightedness is incurable. Amati showed it could be corrected.

Within less than ten years, the invention had spread all over Italy. A monk of Pisa named Alessandro della Spina seems to have purchased a pair and worked out the secret for himself; his tombstone says that he made the method generally known. It quickly spread across France, Germany and the Netherlands; and the Dutch, with their inborn talent for precision, soon became the foremost manufacturers of lenses in Europe.* All this exciting early history is lost, except for a few brief references in

Galileo showing his telescope to the Doge of Venice, by Sabatelli Luigi
Opposite: The discovery in Middelburg of the properties of lenses which led to
the invention of the telescope

manuscripts of the period; yet from the speed with which the invention spread, we can imagine how eagerly Amati's contemporaries seized on this remedy for failing sight; it must have struck them as halfway towards the invention of the elixir of life.

Someone – probably Amati himself – must have noticed that if two lenses are held up, one in front of the other, distant objects can be enlarged. (I recall making the same discovery myself at the age of six or seven in the shop of an uncle who was an optician.) If so, no one thought of making practical use of the discovery. Two and a half centuries passed before a British scientist, Leonard Digges, performed a series of experiments with lenses – clamped on frames – by which he was able to see 'seven myles off' and 'declare what hath been doon at that instance in private places'.[1] Leonard Digges had invented the telescope, although he had not taken the step of placing his lenses at either end of a tube. And although his son Thomas Digges – a pupil of the magician John Dee – became an eminent astronomer, it never seems to have dawned on him that his father's invention would be of use in his observations.

* Because of the difficulty of manufacturing clear glass, these lenses were often made of quartz and beryl.

123

But the time was obviously ripe, and a few years after the turn of the century – when the momentous meeting between Tycho and Kepler took place – several Dutchmen seem to have discovered the telescope simultaneously. The original credit should probably go to some children who were playing with lenses in the town of Middelburg, and noticed that they could make the weathercock on the church appear much closer, although it was also turned upside down. A local optician, Zacharias Janssen, tried putting the lenses into a tube, but felt that the value of his invention was diminished by the fact that it turned things upside down. This, in fact, is what happens if both lenses are convex. Another spectacle-maker of Middelburg, Johan Lippershey, tried putting a concave lens in the end of the tube nearest the eye, and discovered that the image now remained upright, although the magnification was less. In October 1608, Lippershey applied for a thirty-year licence to manufacture his instrument exclusively; the Dutch government was impressed, and asked him if he could make an instrument that could be used with both eyes at the same time. Lippershey invented binoculars during the following month, and the government bought them; but they refused to grant his licence because two other men (Janssen, and a James Metius of Alkmaar) also claimed to have discovered the telescope.

In Italy, the news soon came to the ears of a brilliant and bellicose man of science, Galileo Galilei, the professor of mathematics at Padua University. Galileo was known to the learned world as an inventor, and had his own workshop. The construction of a telescope offered no problems; he had one ready by August, and invited the Venetian senate (Padua belonged to Venice) to come and test his instrument from the tower of St Mark's. This telescope, which magnified things nine times, was a tremendous success. Galileo pointed out that it would enable them to see the arrival of ships hours before they were visible to the naked eye, and be invaluable insurance against attack from the sea. He presented his telescope to the senate, and they were so grateful that they voted to double his salary. By this time, local spectacle-makers were manufacturing telescopes and selling them in the streets. To keep the senators from regretting their bargain, Galileo set about constructing a far more powerful telescope. He soon claimed to have made one that magnified a thousand times.

One fine night in the autumn of 1609, Galileo stepped out of doors and pointed his new telescope at the moon. What he saw left him breathless. In a few seconds, a glance through his telescope destroyed the idea of the moon that had been taken for granted for four thousand years. The ancients had believed that the moon is a smooth, shining globe; Dante expressed the general view when he described the moon in the *Paradiso* as 'lucid, dense, solid and polished, like adamant shining in the sun'.[2] Dante also believed that the moon shines by *its own* light as well as reflecting the sun. The shadows were assumed to be reflections from the earth. Galileo's telescope revealed a globe that looked like a great, corroded sheet of copper, or the face of a man who has suffered from smallpox. On closer examination, it was clear that the holes were not below the surface, but above it, on the tops of immense mountains. A few long, deep score-marks were obviously valleys. Amid the shadows on the surface there were glittering patches of light, which slowly expanded and joined with the larger areas of light. Clearly, the moon was not like a mirror, or an enormous lamp; it was a world.

He pointed his telescope at the stars, and was astounded when hundreds more stars seemed to spring into existence. He looked at the belt of Orion, with its nine stars, and

Galileo's wash drawings of his observations of the moon

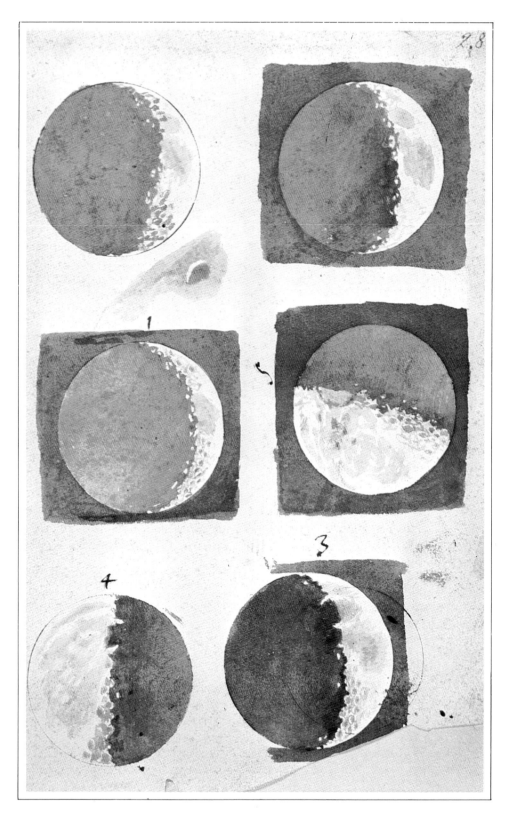

was able to count eighty more. He turned his telescope on the Milky Way, which looks like a band of luminous gas, and the band turned into thousands of stars, 'planted together in clusters'.

The most remarkable discovery was still to come. On 7 January 1610, Galileo turned his telescope on Jupiter, visible to the naked eye as a yellowish star. The lens showed Jupiter as a large globe, slightly flattened at the poles. Galileo's telescope was not powerful enough to show the streaky white bands, caused by cloud belts, or the great Red Spot in its northern hemisphere. But it *did* show him three tiny, bright stars close to the edge of the planet. At first he assumed them to be ordinary stars; yet he was fascinated by their unusual brightness, and by the fact that they were in a straight line

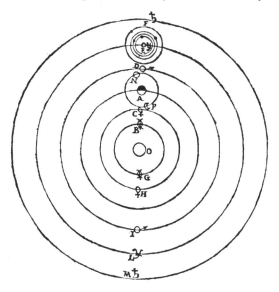

Galileo's drawings of the Copernican system, showing the four satellites of Jupiter

parallel to the planet's ecliptic – its 'waistline'. The following night, when he looked again, he was surprised to see that the 'stars' had moved to the other side of the planet. Further observation revealed a fourth satellite. He had discovered Jupiter's 'moons'.

The significance of this was tremendous. For more than twenty years, Galileo had been convinced of the correctness of the Copernican system – he had admitted as much in a letter to Kepler in 1597. But in the same letter – thanking Kepler for his *Mysterium Cosmographicum* – he also admitted that he was afraid to express his belief openly, in case he met the same fate as Copernicus, 'ridicule and derision'. And so he continued to teach his students the views of Ptolemy. He was aware that one of the main arguments against Copernicus – perhaps the only one that appeals to scientific common sense – is that the moon is quite obviously a satellite of the earth. So is it not rather absurd to believe that the moon travels round the earth, while the earth – and the other planets – travel round the sun? Why should it be an exception to the rule?

The discovery of the moons of Jupiter meant that our moon was no exception. The chief objection to the Copernican system had suddenly dissolved. Shortly afterwards, Galileo's last misgivings vanished when he turned his telescope on Saturn, the outermost planet, and saw what he took to be two moons on either side of it. Saturn *has* moons – ten of them – but what Galileo saw were fairly certainly its 'rings'.

Galileo was forty-five years old at the time he made his discovery, and he had so far published nothing – he had preferred to circulate his writings in manuscript. But these revelations were clearly another matter. Surely, even the most bigoted churchmen could now no longer ignore the truth of Copernicus? Possessed by this belief, in a state of exultant excitement, Galileo wrote a booklet called *Sidereus Nuncius*, *The Messenger from the Stars*; it was published in Venice in March 1610, less than two months after he had seen the moons of Jupiter. And the effect was certainly very much what he had anticipated. The work caused an uproar, and its fame spread quickly across Europe; Kepler heard about it within a matter of weeks. It was one of those books that creates feverish curiosity, an immediate desire to read and discuss. We may find this difficult to understand, since it was far less epoch-making than Copernicus's *On the Revolutions* or Kepler's *The New Astronomy*, both of which had failed to excite much attention. But then, Galileo's work was about practical matters, not theories; it was almost a kind of travel book, and its impact could be compared to that of Heyerdahl's *Kon-Tiki Expedition* after the Second World War. It was about the depths of space and about other worlds. And its implications were as important as its actual arguments. If space was full of millions of invisible stars, then the argument that the stars were made to light up the night sky was obviously shaky. In fact, Galileo finally brought home to his contemporaries what Aristarchus had tried to convey to the Alexandrians nineteen hundred years earlier: the vastness of the universe. Even Kepler was shaken; he kept repeating: 'The infinite is unthinkable. . . .'*

At the age of forty-five, Galileo had finally achieved what he always felt he deserved: European fame. He had always been rather jealous of his younger contemporary Kepler – as is evidenced in his grudging attitude to the latter's great discoveries. Now, with the aid of the telescope and a single short book, he had become more famous than Kepler. There are still many educated people who believe that Galileo invented the telescope. The irony is that he was not even the first to use it to look at the moon. The precedent here, as in the matter of its invention, goes to an Englishman. In July 1609, Thomas Harriot, a friend of Sir Walter Raleigh, looked at the moon through a telescope of his own construction, and saw what Galileo had seen. When he finally received a copy of *The Messenger from the Stars*, he was unimpressed with Galileo's sketch of the moon, and made a far more accurate one of his own, containing no less than fifty craters and valleys. As we recall, opticians in Venice were selling telescopes – to Galileo's embarrassment – within weeks of the demonstration from St Mark's. But fate had chosen Galileo to be the foremost representative of the 'new ideas' – and to be the chief target of the rage and alarm they aroused. For the remaining thirty-two years of his life, Galileo would be in the front line of the 'warfare between science and theology' (to quote the title of a famous nineteenth-century book on the subject). As a result, his name has become a symbol of scientific integrity faced with superstition, bigotry and malice.

In fact, this view of the matter is crudely oversimplified. It is important to understand why Galileo became the focus of so much hostility: because, unlike Copernicus and Kepler, he was aggressive, conceited and sarcastic. From his earliest days, he seems to have had a talent for rubbing people up the wrong way. The charitable view – which, in view of his genius and later sufferings, it is easy to take – is that his brilliance and vitality made it difficult for him to suffer fools. But there is no evidence that he tried.

Galileo seems to have inherited his contempt for authority from his father Vincento,

* But this was because he felt, quite rightly, that the universe cannot contain an infinitude of stars.

a musical historian and composer, noted for the vigour of his polemics and his sharp wit. The family were impoverished nobility, and there was not enough money to keep Galileo at the medical school of the University of Pisa. His application for a scholarship was turned down. Yet he had already given proof of a remarkable intellect, and there were forty scholarships available. That he failed to obtain one suggests that he had been exercising his talent for making himself unpopular.

The event that gave direction to Galileo's life occurred at the age of eighteen. Seated in the cathedral at Pisa, he observed the swings of a lamp suspended from the ceiling, and took the trouble to time them. As a result, he discovered that the long swings and short swings took precisely the same time. His father had refused him permission to study mathematics – feeling that he would do better to concentrate on diseases; but Galileo sat at the back of a geometry lecture, and decided that science and mathematics interested him more than medical problems. After leaving the University without a degree – at the age of twenty-one – he continued his studies in mathematics at home, and wrote a treatise on hydrostatics that impressed a scholarly marchese, Guidobaldo del Monte. As a result of the patronage of the Duke of Tuscany, Ferdinand de Medici, he was made a lecturer in mathematics at the University of Pisa at twenty-five.

During the next three years, Galileo gave further proof of his experimental genius when he climbed the leaning tower of Pisa and simultaneously dropped a heavy and a light cannon ball; they struck the ground at the same moment, disproving Aristotle's assertion that heavy bodies fall faster than light ones. Legend has it that he later invited the whole faculty to witness the refutation of Aristotle. This is unlikely, since the majority of his colleagues seem to have been hostile to him, and Galileo aggravated the situation by circulating satirical poems about his critics. The trouble here may have been the outspokenness of his contempt for Aristotle. But Galileo's Russian biographer Boris Kuznetsov puts his finger on the real problem when he says: 'For Galileo, the University of Pisa was not too Aristotelian – just too provincial.'[3] So when, three years later, his kindly patron obtained him a professorship at the University of Padua, his departure caused general relief.

Here he worked on quietly for the next eighteen years – until he made his first telescope – building up a reputation as a mathematician and inventor. (He invented early versions of the thermometer and metronome.) Yet he made no attempt to publish his major discoveries – neither the law of oscillation, nor the law of falling bodies. He also saved for his private correspondence what we now regard as his most important achievement: the first law of motion, which states that a body will continue to move in a straight line until it is acted on by some outside force. This sounds unexciting enough – even rather obvious – yet for that time it was a revolutionary advance. Aristotle had explained that bodies move solely because they are *pushed*, and when the push ceases, they stop. So why does a javelin fly through the air even after the thrower's hand has stopped 'pushing' it? Because it has created a vortex in the air, and the air continues to push it forward. (He makes it sound like jet propulsion.) But Galileo disproved this by a simple experiment. He rolled a weight down an inclined plane, and allowed it to roll up another plane on the other side. He noted that the weight would roll as far up the second plane as it had rolled down the first, which implied that a moving body would continue to move unless something stopped it.

Now we may recall that Kepler was bothered by the problem of what caused the

Seventeenth-century armillary sphere, or skeleton celestial globe

planets to move, and invented the idea of a force that emanates from the sun, creating a kind of vortex (a good Aristotelian notion). But Galileo's discovery implied that a 'force' or whirlpool was unnecessary. Once the planets were moving, they would keep moving until something stopped them.

All that Galileo needed to complete his picture of the solar system was the force of gravity, to *attract* the planets (instead of pushing them) and cause them to move in a circle. But this important insight eluded him. So his system still lacked that one vital component that might have overwhelmed the sceptics. As it was, his opponents could say: 'Very well, suppose we concede that a moving body will keep on moving unless something stops it. But in that case, why do the planets not keep moving in a straight line, into the depths of space?' And to this, Galileo had no answer.

In 1609, the year he made his first telescope, his theory of motion was known only to a few correspondents. But with his clear admission, in *The Messenger from the Stars*, that he supported the Copernican system, the years of quiet neutrality were over. He was in the firing-line. And, what was worse, he was widely detested, so that his opponents had an additional reason for dismissing his discoveries. Arthur Koestler, describing Galileo, speaks of the 'cold, unrelenting hostility which genius plus arrogance minus humility creates among mediocrities'.[4] But when we look closely at Galileo's life, even this explanation seems inadequate to explain the fury he aroused.

What we need here, I would suggest,* is the psychological theory of the 'Right Man', advanced by the writer A. E. Van Vogt. Van Vogt points out that there is a certain type of man whose sense of security is threatened by the notion that he might be wrong or mistaken. Being infallible is an important part of his image of himself. His emotions are under inadequate control, and he explains the dislike he arouses in others by telling himself that people cannot bear the fact that he is right and they are wrong.

This is, of course, a residue of childishness – Freud remarks somewhere that an angry child would destroy the world if it had the power. In most people, it is modified by realism – which is another name for caution. Oddly enough, the people in whom it remains unmodified are often those with the greatest potential. Their feeling that they possess unrecognized capacities strengthens their belief in their innate 'rightness'. If such people achieve any kind of authority, they are likely to become tyrants, for they find it quite impossible to distinguish between just and unjust indignation; anything that arouses their anger demands instant punishment.

Galileo was spoilt; he grew up with a chip on his shoulder. When the University of Pisa refused him a scholarship, he knew they were wrong. When distinguished patrons like del Monte and Medici recognized his merit, it confirmed his sense that really intelligent people knew genius when they saw it. Then followed the eighteen years of relative obscurity at Padua, during which he kept his Copernican views a secret. Why? He was in no danger of persecution. But the 'Right Man' is afraid of nothing so much as ridicule. It touches a secret spring of paranoia, and drives him to frenzy.

Without this recognition, it is impossible to understand why Galileo's open acknowledgement of his Copernican views aroused such passions. After all, the Jesuits themselves were enlightened humanists, and some of them were already inclined to accept the Copernican theory. Father Clavius, chief astronomer of the Jesuit College in Rome, confirmed the existence of the satellites of Jupiter and Saturn, and also con-

* For a fuller account, see my *Mysteries* (1978), Part 2, chapter 1.

Galileo the beleaguered sage. A conventional view by Tito Lessi

firmed Galileo's observation that Venus has phases, like the moon – which proved that it must revolve round the sun. But this does not mean that the Jesuits were full-blown Copernicans. They could see that Ptolemy was wrong, and that therefore Copernicus's criticisms were correct. But then, there was an alternative system – Tycho's – which still placed the earth at the centre of the universe. Tycho had worked out for himself the same system as Heraclides, in which Mercury and Venus circle the sun, and the sun circles the earth. So the Jesuits were able to reject Ptolemy, while preserving an open mind about the true arrangement of the solar system.

Preserving an open mind was precisely what Galileo found impossible. Once he was convinced he was correct, he had to say so, loud and clear. If challenged, he would, no doubt, have defended himself by saying that he was a scientist, and that it was his business to tell the truth. Yet in retrospect, it is hard not to feel that his real concern was that *he*, Galileo, was correct, and that the world ought to know this and pay him due homage.

In fact, he received a great deal of homage in the early days after his discovery. The Academy of Science made him a member and gave a banquet in his honour (in which the president, Prince Federico Cesi, christened the new invention 'the telescope'). The Pope gave him a friendly audience, and the Jesuit College performed various ceremonies in his honour. It was true that his academic colleagues – who had a vested interest in Aristotle – were almost uniformly against his discovery, and that some even declined to see the moons of Jupiter when invited to peer down his telescope. But

Galileo's supporters were the most influential men in Italy. (He even tried to name the four satellites of Jupiter after children of Ferdinand de Medici, but the rest of Europe preferred Greek classical names like Io and Ganymede.)

And at this point, Galileo's conceit began to turn the tide against him. That it had not done so earlier is a matter for some surprise; for example, he had treated Kepler with scant courtesy even while begging for his support. But he now made the mistake of antagonizing a Jesuit. In 1612, the assistant of Father Scheiner, a Jesuit astronomer at Ingolstadt, turned his telescope on the sun and saw sunspots. He published his observation, and a copy of his report went to Kepler and Galileo. Kepler replied that he had once seen something of the sort with his naked eye, and had mistaken it for Mercury. Galileo delayed replying for three months, then declared that *he* had discovered sunspots in 1610. He went on to write a short book about them, in which he again declared his support for Copernicus. Still no churchman raised objections; but Father Scheiner was understandably antagonized by the rudeness – he might have been justified in asking Galileo why he had kept the sunspots to himself for two years.

In Pisa, the professor of mathematics was Father Castelli, a former pupil of Galileo. In 1613 he had been invited to dine at court, where a controversy arose about the satellites of Jupiter and the Copernican system. Castelli defended it against the accusation that it was contrary to holy scripture, and wrote Galileo an account of the argument. Galileo experienced his usual sacred rage at being criticized, and dashed off a pamphlet, the *Letter to Castelli*, in which he broke the unwritten law, and waded knee deep into theological controversy. His opponents must have chuckled and rubbed their hands. At last – as anyone might have predicted – he was doing their job for them. Theology, said Galileo, is the Queen of the Sciences because she deals with matters of spiritual revelation; so in matters of physical science, let her mind her own business. He put it less bluntly than that, but the effect was much the same.

A year went by without repercussions; then a certain aged Father Lorini, with whom Galileo had once picked a quarrel, saw the pamphlet, and was outraged. He wrote an indignant letter to the Holy Office in Rome – the Inquisition – quoting (and misquoting) the *Letter to Castelli*. (For example, where Galileo had written that certain scriptural passages, taken in the literal sense, 'look as if they differ from truth', Lorini quoted him as saying they were 'false in the literal sense'.) The Holy Office examined the pamphlet, and decided that it contained nothing actually contrary to Catholic faith. Again, Galileo was saved. But his friends in the Church now began to have their doubts. And once the controversy had been set rolling, it was difficult to stop. A Father Foscarini wrote a book defending the opinions of Copernicus and Galileo, and it was sent to Cardinal Roberto Bellarmine (subsequently canonized) for his opinion; Bellarmine – who had been involved in the burning of Giordano Bruno for heresy sixteen years before – again reacted with good sense and reason, saying that there was no objection whatever to the presentation of the Copernican system as an interesting, and possibly true, hypothesis, but that it must not be asserted as out-and-out fact. He said that if Galileo thought it was true, then it was up to him to prove it. And this, of course, was precisely what Galileo could not do, since he lacked that essential component of proof, a theory of gravitation. Galileo blustered and grunted and grumbled, and pretended that his opponents were all stupid and prejudiced Aristotelians – which was now manifestly untrue. In *The Sleepwalkers*, Koestler remarks percipiently: 'He had committed himself to an opinion, and he must be proved right; the heliocentric system had become a matter of personal prestige.'[5] Finally, in February 1616, the Holy

Office came out into the open, and declared that the Copernican system was untrue, and must not be taught any further 'until corrected'. Again, the wording left room for manoeuvre. In fact, Copernicus's book was removed from the index a mere four years later, in 1620, with a few sentences removed. The Church was leaning over backwards to be reasonable.

But Galileo was not. No one had openly attacked him – his name had been tactfully ignored – but he felt that his enemies must be rubbing their hands, and silently vowed to have the last laugh. Instead of thanking his stars for his escape, he returned to Florence – he had left Padua in 1611 – and brooded angrily for the next seven years. Somewhere in his unconscious mind he was nagged by the problem of gravity; for he spent much time on a theory of the tides. But instead of recognizing the moon as their cause, he believed it was the rotation of the earth. He wrote a bitter and bad-tempered book called *Il Saggiatore* (*The Examiner* – of precious metals) in which he attacked, among others, Tycho Brahe. He dedicated the book to the new pope, Urban VIII, Maffeo Barberini, who had attempted to intervene in his favour in 1616. Now that Barberini was Pope, Galileo was convinced that his troubles were over. When he visited Rome, Barberini showed himself just as gracious as Galileo had expected, giving him expensive presents and conferring a pension on his son. But where the Copernican system was concerned, the Pope refused to budge. It could be taught as hypothesis, he said, but not as fact. And if Galileo meant to write a book about it, he must make that quite clear. Galileo agreed sulkily and returned to Florence. He wasted the next five years writing a book on his fallacious theory of the tides. But in January 1630, he completed his *Dialogue on the Two Chief World Systems*. And, inevitably, Galileo being the man he was, it was a betrayal of the promise he had made to the Pope. The *Dialogue* is a straightforward argument in favour of the Copernican system.

At this stage, the Pope was still unaware of this. A few months after the work's completion he again received Galileo in friendly audience, and again emphasized that he had no objection to a book on the Copernican system, provided it was taught as a theory, not as proven fact. Galileo appeared to agree; and in due course, submitted his book to the chief censor and licenser, Father Riccardi, who would have to give it his *imprimatur*. This good natured priest found it all above his head, and asked his assistant to look at it. They seem to have realized that it was not quite as 'hypothetical' as it should have been, and made a few minor adjustments. But after all, Galileo was a friend of the Pope, and the Pope seemed to have approved of his book. So after taking advice from various other scholarly clerics, he allowed it to be printed. It appeared in February 1632.

At this point, finally, the Pope read it, and exploded. This was not simply disobedience; it was outright defiance. Galileo had not only ignored the Pope's advice; he had even taken certain suggestions made by the Pope, and put them into the mouth of a person (called Simplicio) who was obviously intended to be an idiot.

But Galileo had mistaken his man. Whatever Barberini's virtues might be – and he was a man of considerable intellect, vitality and charm – he was also another Right Man. Jacob Bronowski has an apt description of him in his *Ascent of Man*: 'He had a confident, impatient turn of mind: "I know better than all the cardinals put together . . ." he said imperiously. But in fact, Barberini as Pope turned out to be pure baroque: a lavish nepotist, extravagant, domineering, restless in his schemes, and absolutely tone-deaf to the ideas of others. He even had the birds killed in the Vatican gardens because they disturbed him.'[6] A man like that would not allow anyone to treat him like a fool. The book was confiscated and Galileo was summoned to Rome to appear before the Inquisition.

The trial of Galileo is usually represented as the ordeal of an unworldly scientist at the hands of a pack of sadistic bigots. We have seen that this view is something like an inversion of the truth. Galileo was not tried for believing in Copernicus; he was not even tried for writing about his belief in Copernicus. He was really tried for dishonesty, trickery and for breaking his word. And had the Pope framed the indictment, he would no doubt have added: for ingratitude. For Galileo had been treated with the utmost warmth, confidence and consideration. There had been no question of bullying or threats. He had been asked, as a favour, not to embarrass the Church by stirring up conflict, and he had promised to comply. All that was required of him was that he state clearly that the Copernican theory *was* theory, not a proven fact. (Even today,

A mariner determining his position on dry land with an early navigational instrument

when it *is* proven, we still refer to it as the Copernican 'hypothesis'.) This was no more than the truth, for Galileo could *not* prove it. There was a gaping hole in his theory – which Newton would fill with the concept of gravity – and Galileo was too much of a self-deceiver to admit it.

But then, Galileo had calculated shrewdly that he would get away with his dishonesty. And he was right. The Pope might rage and stamp, but there was little he could do. In Italy, burning heretics was almost unheard of; even Giordano Bruno would have escaped the stake if he had merely acknowledged that his views were mistaken. Burning – or even imprisoning – a famous scientist like Galileo would have subjected the Church to the denunciation of every humanist scholar in the world, whether Catholic, Protestant or Mohammedan.

On 12 April 1633, Galileo appeared before the Inquisition. He had almost nothing to

say in his own defence. He admitted that in 1616 he had accepted the instruction not to hold or teach the opinions of Copernicus, and he admitted that he had not asked permission to write the book which broke that prohibition. He also admitted that when submitting the book for its *imprimatur*, he had kept quiet about his promise. It was an open-and-shut case, and the Inquisition would have been justified in sentencing him to prison. But the court did not want to force the issue, and it did not meet again. Galileo was merely told that he had to retract his assertion that the sun was the centre of the universe, and he did so. Legend has it that he added under his breath, '*E pur si muove*' ('It moves all the same'). The legend was almost certainly started by Galileo himself. Just as he had expected, he had escaped scot-free.

Having made this retraction, he was allowed to return home. No doubt feeling that he had been utterly outwitted, the Pope expressed his annoyance by placing Galileo under house-arrest, and ordering him not to write any more. It made no real difference; Galileo retired to his farm and went on much as before, baiting his enemies, and writing a book called *Two New Sciences*. This was carried off by the French ambassador and printed in Holland; but no one attempted to call Galileo to account. He died in 1642, at the age of seventy-eight, and the Grand Duke of Tuscany was only prevented from erecting a monument to him by direct prohibition of the Pope. Galileo had not only won; he had done so in such a way that his enemies – or rather, the friends he had forced into that position – were made to look as if they were the guilty party.

In provoking this completely unnecessary conflict, Galileo had performed an enormous disservice both to the Church and to science. The Pope had no wish for a head-on collision with the humanists; but once Galileo had thrown down the gaunt-

Milton visiting Galileo in prison. The myth of Galileo's martyrdom was soon widespread

let, he was not one to retreat. The scientific tradition in Italy came to an abrupt halt, and the initiative passed to the northern lands. Galileo's trial had placed the Church in a completely false position, making it look as if science stood for honest investigation while the Church stood for superstition and stupidity. Before Galileo, there had been no 'warfare between science and theology'. It was he who provoked hostilities.

But we are also in a position to understand the deeper implications of what had happened. In effect, Galileo had taken a hatchet and sliced the bicameral mind in two. Kepler, whose contribution to astronomy was far more important than Galileo's, was a natural 'unificationist'. He felt about the universe exactly as the ancient Egyptians had – that there were somehow invisible threads connecting man with the stars. The Church agreed with him. Galileo, by acting as *agent provocateur*, had also forced science into a position of rigid 'objectivism' – that is, materialism.

We have seen the fundamental paradox involved in 'bicameralism'. The Egyptians may have been wrong to believe that the earth was the centre of the universe; yet that belief arose out of their perception that there *is* a connection between man, the earth and the planets. If the Greeks rejected Aristarchus's 'Copernican theory' because it frightened them, it was also because they were trying to hold on to that same intuitive perception. What the human race needed was time; and with the coming of the Dark Ages, it had more than enough. In spite of the Counter-Reformation, the Church showed considerable common sense in adjusting to the new theories of science. Even the most anti-clerical historians find it hard to disagree that, without Galileo, the Church would have come to terms with Copernicus by the end of the seventeenth century. But after the trial of Galileo, the time for peaceful compromise was past. The Church was forced into the position of the enemy of science, a position it would continue to maintain – in zoology, geology and biology – for the next two and a half centuries. That split divided intelligent human beings into two camps: the religious men, who continued to believe that, in a basic sense, man is still the centre of the universe; and the scientists, who asserted that man is merely an insect who happens to inhabit a second-rate planet in a third-rate solar system in a fourth-rate galaxy.

Meanwhile, the astronomers were quite unconcerned with the earth's relative status in the universal scheme of things. They continued to be absorbed by the tremendous panorama opened up for them by the telescope.

While Europe was being torn apart by the Thirty Years' War and the Roman Catholic Church was wasting its strength in its struggles with various forms of heresy – from Protestantism and Jansenism to the Copernican theory – British science and British trade were benefiting from England's isolation from the continent. The defeat of the Spanish Armada in 1588 brought her freedom of the seas. But navigation remained a major problem. To begin with, the magnetic compass seemed unreliable. Columbus was one of the first to notice this, on his first voyage across the Atlantic in 1492; in the Caribbean, there was dismay when the needle no longer pointed due north.

In 1600 this problem was solved by a wealthy physician named William Gilbert, in a series of brilliant experiments with magnets. His treatise *De Magnete, Magneticisque Corporibus* exerted an immediate and tremendous influence. Gilbert suggested that the earth was an enormous magnet, with a north and south pole – which explained why the compass often seemed to deviate from true north. He also devoted a chapter to demolishing Tycho Brahe's theory that the earth stands still, demonstrating with cogent arguments that it had to revolve on its axis.

This led him to another interesting problem. If the earth rotated once a day, why did things stay on its surface, instead of flying off into space? Gilbert's answer was that the earth's magnetism also explained this anomaly; it held objects down by magnetic attraction. He was wrong, of course; but it was one of those mistakes that is ten times more productive than any number of correct theories. For before Gilbert, it had never occurred to anyone to wonder why objects stuck to the surface of the earth; it was assumed that things had 'weight', and that was all there was to it. By questioning this Gilbert prepared the way for the theory of gravitation. He went even further, and suggested that this same magnetism prevented the earth's atmosphere from flying off into space. Again, the suggestion was breath-taking. The ancients had always assumed that the universe is full of air; Aristotle had said that a vacuum was impossible. Now suddenly, Gilbert was suggesting that the atmosphere had a certain depth, like the sea. For the past fifty years or so, miners had been aware of the curious fact that water could not be pumped out of mines more than thirty-two feet deep. And in 1630, a pupil of Galileo named Torricelli would connect this fact with Gilbert's theory about the atmosphere, and realize that the weight of the atmosphere at any given point is equivalent to the weight of a column of water thirty-two feet high.

By 1642, the year Galileo died, European science was in a state that could be described either as intellectual ferment or as intellectual chaos. The great French mathematician Descartes decided to suppress his treatise, *Le Monde*, which supported the Copernican doctrine – not because he was afraid of persecution, but because he was a faithful son of the Church, yet was equally certain that Copernicus was correct.

*Above: Gilbert's diagram
showing the earth's centre as an
enormous magnet
Right: William Gilbert,
author of* De Magnete

Descartes had argued that even if God had created any number of distinct universes, they would all have to be governed by natural law. And here were Galileo's opponents pointing out, quite rightly, that natural law failed to explain why the planets moved in circles, let alone *how* they could move in a vacuum. Science knew more than ever before; yet every new answer raised another dozen questions.

The man who brought order into this chaos was born in Lincolnshire on Christmas Day, 1642, the year of Galileo's death. Isaac Newton's mother had been severely depressed by the death of her husband, and her baby was born prematurely. The women who were sent to the village for medicine expected him to die before they returned. But the sickly child slowly improved under his mother's devoted care. He was to grow up with a deep, almost morbid, attachment to his mother. He was to lose her at the age of four – not through death, but through marriage; she went to live with her new husband, the Reverend Barnabas Smith, in his parish at North Witham. Newton was left in the charge of his grandmother at Woolsthorpe. He was never able to throw off a kind of emotional constipation.

Like Kepler, Newton was a delicate child. His inventive genius was apparent from an early age – he made a clock, a water-wheel, and a mill that ground flour (powered by a mouse). Otherwise, he seems to have been a fairly normal boy; he read a great deal and, when sent to lodge with the local apothecary, fell in love with his stepdaughter, a Miss Storey, and became engaged to her. His mother's second husband had also died, and Newton showed himself efficient and practical in helping her to run the farm at Woolsthorpe. He seemed scheduled for a fairly normal career as the local squire

Above: Descartes' idea of terrestrial
and celestial magnetism
Left: René Descartes, philosopher
and methodologist of science

(Woolsthorpe was a manor house). But a maternal uncle, the Reverend William Ayscough, noted the boy's natural intelligence, and persuaded his mother that he ought to be sent to university. As a result, he became a sizar at Trinity College, Cambridge, at the age of nineteen. A sizar was a student who helped to earn his keep through menial services – Newton's family was not wealthy. This seems to explain why the engagement to Miss Storey was allowed to lapse – she was also penniless, and their future together looked unpromising.

During his first three years at Cambridge, 1661–4, Newton seems to have shown no

Isaac Newton, Cambridge's young genius

unusual promise, although he showed himself a diligent student of mathematics under his professor, Isaac Barrow. Intellectual life at Cambridge was not particularly lively; the Civil War had taken its toll, funds were short, and there was an atmosphere of fatigue and apathy. Then, in 1664, the plague came to Cambridge; and the University had to close down altogether.

It was the best thing that could have happened to Newton. In old age, he wrote about his early work: 'All this was in the two plague years of 1665 and 1666, for in those days I was in the prime of my invention, and minded mathematics and philosophy more than at any time since.'[7] And one of the problems to which he was devoting his mind was the question of what made the moon and planets move in circles (or ellipses). The famous story tells how he was sitting under an apple tree during this enforced retirement from Cambridge, when he was aroused from his meditations by the fall of an apple. (Voltaire was told the story by Newton's niece, Catherine Barton, who lived with him for many years.) He had been brooding on Kepler's laws of planetary motion, wondering what could account for their circular path. Now, in a flash, he saw the answer: the same 'magnetism' that pulled the apple towards the earth. But in spite of its 'inspirational' nature, it is doubtful whether Newton would have conceived the universal law of gravity without knowing about Gilbert's magnetic hypothesis.

But what is the law that governs this attraction? Here again, Newton had been handed the clue by a predecessor – Kepler, whose third law stated that the cube of the planet's distance is equal to the square of the time it takes to revolve. Newton calculated what force would keep the moon in its orbit around the earth. Obviously, it must weaken inversely with the distance – that is, the greater the distance, the smaller the force. He soon reached the conclusion that the force must weaken inversely as the square of the distance. (That is to say, if the distance doubles from one unit to two, then the force becomes $2 \times 2 = 4$; but since the force varies *inversely*, this becomes $\frac{1}{4}$. If the distance trebles, the force becomes $\frac{1}{9}$, and so on.) Newton later admitted that he had seen so far because he 'had stood on the shoulders of giants'.[8] And the most important of these giants was Kepler.

Having found his solution, and worked out that the orbit of a planet was bound to be an ellipse, Newton pushed it aside and forgot about it. This was not because he thought it unimportant, but because his mind was occupied with many other matters. To begin with, it was difficult to work out problems involving the changing speeds of the planets, or the problems involving a new kind of mathematics. The old algebra and trigonometry dealt with unchanging quantities. But suppose a problem involves changing speeds? Or has two variables? Suppose, for example, that you are a manu-facturer of tin cups, and you want to know how wide and high a cup should be to give the largest possible volume for the smallest amount of tin plate?* You might solve the problem by simply trying various combinations. But is there not a less troublesome method that will tell you the *rate of change* of volume compared to height and diameter? Newton evolved such a method, which he called 'fluxions' (and which we now call the infinitesimal calculus), which would apply equally well to the volume of a tin cup or the movements of the moon. It was the greatest mathematical advance since the Greeks; and, typically, Newton could not even be bothered to publish it. (Years later, this led to bitter controversy when the German philosopher and mathematician Gottfried Wilhelm von Leibniz rediscovered the calculus.)

The third major problem to which Newton applied his mind in the two 'plague years' was the nature of light. Here again, it was astronomy that set him thinking about the problem. As everyone now knows, a prism 'breaks up' white light into its constituent colours. The edge of a lens is, of course, a prism. And astronomers of the seventeenth century found that the image of a star or planet was often a confused blur of colours when seen through their refracting telescopes. They found that this irritat-ing phenomenon – called chromatic aberration – became no worse if the telescope was made far longer; but the longer the telescope, the higher the magnification. Their solution was to make them longer and longer. The German astronomer Johannes Hevel (called Hevelius) – famous for his atlas of the moon – built a telescope 150 feet long, which had to be suspended from a flagpole with many ropes. Sir Christopher Wren was then in the process of building St Paul's Cathedral, and English astrono-mers wondered whether they could not use its scaffolding for suspending an even

* The answer is: when the height and diameter are equal.

141

larger telescope. . . . But the whole approach was obviously unsatisfactory. This was why Newton began his investigation of light by asking a simple question: was the glass itself responsible for causing the colours of the spectrum? If so, then the problem was simply to find a purer form of glass.

Newton investigating light. An imaginative reconstruction by J. A. Houston, RSA
Previous page: William Blake's painting of Newton the rationalist

He devised a marvellously simple experiment to find out. A beam of light was allowed in through a hole in his blind, and allowed to pass through a prism. This cast the usual spectrum of colours (Newton invented the word spectrum) on a board. Newton now made a small hole in the board, so that only one colour could shine through it. Then he set up another prism on the other side of the board, and allowed the coloured light to pass through it. The second prism failed to change the colour of the light – proving that it was not an impurity in the glass that caused colours. Light itself was made up of the colours of the rainbow. He made the result doubly certain by reversing the experiment – focusing the seven colours of the spectrum on to a single point – whereupon they turned into white light.

Newton jumped to the incorrect conclusion that chromatic aberration cannot be eliminated; but the error was fruitful in that it led him to ponder other methods of focusing the light in a telescope. In fact, a Scots mathematician, James Gregory, had worked out the answer two years before the plague. A curved mirror also has the power to focus light and magnify an image – as the distorting mirrors at a fairground demonstrate. And since the light bounces straight off the surface of a metal mirror, and does not have to pass through glass, this seemed the obvious solution. Newton read Gregory's paper, and employed his leisure at Woolsthorpe constructing a small reflecting telescope. He peered through it at Venus and Jupiter, and verified that there was now no chromatic aberration. His $6\frac{1}{4}$-inch model was the predecessor of the great astronomical telescopes that would reveal new stars and planets.

In 1667 the plague was over, and Newton returned to Cambridge. He was twenty-four, and had laid the foundation of his life's work; he would make no more major discoveries in science or mathematics. His attitude towards his discoveries remained casual – perhaps because he was more interested in ideas than in the fame they might bring. Two years later, his professor, Isaac Barrow, happened to be discussing some problem regarding rate of change, and Newton said that he had worked out a general method for such problems. Barrow asked to see it, and Newton dug out his notes. Barrow's reaction to this new mathematics does not seem to have been recorded; but he persuaded Newton to expand his notes into an essay, which was then copied and sent to various mathematicians. It had to wait another forty years before it was published.

At this time Barrow was anxious to retire from his professorship and return to the subject he preferred – theology. Now he had no doubt about who should be appointed his successor. At the age of twenty-seven, Isaac Newton became Lucasian professor of mathematics in Cambridge. Three years later, he was elected a Fellow of the Royal Society.

During the next few years, Newton applied himself to the problems of optics. Then he learned that a Frenchman, Jean Picard, had measured the length of a degree of longitude, and discovered it to be slightly larger than had been thought – which increased the size of the earth. Again, he turned to his old calculations about the moon, and discovered that the new figures fitted its size and speed and distance from the earth. Now it was obvious to him that his earlier calculations were correct; he had discovered the secret of the 'invisible string' that kept the planets swinging around the sun. Or rather, to be accurate, it was not so much a string as a kind of roundabout. If a cannon-ball is fired at some distant object, the ball gradually falls towards the earth, because the force pulling it down is greater than the force driving it along. Similarly, if a gigantic cannon-ball was hurtling through space, and it came too close to a planet or star, it would be dragged down towards it. But if it was far enough away, and travelling fast enough, it would not fall towards the planet, but would try to rush past it; as it did so, the gravity of the planet would bend its path into a circle. The ball's forward momentum would keep it moving in this circle – until, eventually, it lost sufficient energy to crash down on to the planet (as our moon will one day crash into the earth). This was the vision of gravity that dawned on Newton. He was unaware, of course, that *all* bodies in space are cannon-balls, rushing outwards from some tremendous original bang; but even this idea was implicit in his theory.

And still he made no attempt to publish. The reason, it is now fairly certain, was his increasing interest in the subject of alchemy. This was no casual interest; in fact, Newton preferred alchemy to mathematics (which depressed him if he was forced to

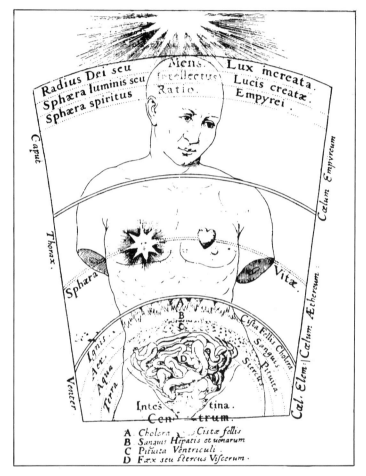

Robert Fludd's cabalist-alchemical view of man as a microcosm of the universe. Newton regarded alchemy as a bridge between the scientific and the spiritual

devote too much time to it). His library contained many books about the secret mystical brotherhood called the Rosicrucians. (One modern researcher, Henry Lincoln, has even found evidence which suggests that Newton was, in later life, the Grand Master of a secret Rosicrucian order.) He certainly believed in the existence of the 'philosophers' stone' which could turn base metals into gold, and spent much of his life searching for it. He even studied the works of the Protestant mystic Jacob Boehme. A century after Newton's death, the poet William Blake used his name as a symbol for the dry, rational intellect which ignores the underlying mystical reality of nature. So it is important to bear in mind that the real Isaac Newton was as much a mystic as Johannes Kepler – or as Blake himself.

One day in 1684, when Newton was forty-one, his friend Edmund Halley called to see him. He wanted Newton's help in solving a dispute. In January of that year, Halley, Sir Christopher Wren and Robert Hooke – a physicist of considerable brilliance, whose character was marred by envy and meanness – had fallen into a discussion about the force between the planets. Halley suggested that it was inversely proportional to the square of the distance; Hooke said he had already reached that conclusion – and had, in fact, proved it mathematically. Wren thereupon offered an expensive book – costing forty shillings – to the first person who could produce such a proof. Apparently Hooke's claim was largely wishful thinking, for by August, he was still no nearer to winning the prize.

What, asked Halley, did Newton think about it? For example, what path would the planets pursue if the inverse square law was correct? Newton answered promptly: an ellipse. How did he know? 'Why, I have calculated it. . . .' Halley asked if he could see the calculation. Unable immediately to find it, Newton promised to send it later. Halley knew that Newton, unlike Hooke, was no wishful thinker; so at the next meeting of the Royal Society he announced that Newton had promised to send his mathematical proof. There was considerable excitement; and the Society seems to have agreed that, when the proof arrived, it ought to be published at the Society's expense.

So the unwilling Newton gave up his attempts to turn lead into gold, and settled down to writing out his proof. It was soon clear that he was not going to be able to launch himself straight into the mathematics of the solar system. First of all, he needed to explain the general laws of all motion – that is, of dynamics. This turned into a work as long as Euclid's *Elements of Geometry*, beginning with the famous three laws of motion: that a moving body continues to move in a straight line unless something acts upon it; that the change of motion is in proportion to the force that causes it; and that when two bodies impinge on one another, their interaction is equal and opposite. (So the earth is pulled as much towards the moon as the moon is towards the earth.)

Having completed his first book of the *Principia* – in one long and gruelling effort – he decided that he now had to refute a suggestion by Descartes – in *Le Monde* – that the planets are dragged around the sun in a kind of whirlpool of particles; so he proceeded to write a second section on the motion of bodies in various media – like a ship in water or an arrow in the air. By this time, Robert Hooke had publicly declared that Newton had stolen his inverse square law from him, since he had told Newton about it six years earlier. (He was unaware that Newton had worked it out for himself more than twenty years before.) Newton lost his temper and declared that he would leave the third book of the *Principia* unwritten. But Halley was persuasive, and the great third book – on the motions of the heavenly bodies – was finally completed in September 1687, just three years after Halley had called on Newton with the problem.

There was still a further embarrassment. The Society was out of funds. Halley generously decided to meet the cost of publication – he was far from being a rich man – and the book appeared in the autumn of 1687, with the name of the diarist Samuel Pepys on the title-page as printer. (He was President of the Royal Society.)

The *Philosophiae Naturalis Principia Mathematica* (*The Mathematical Principles of Natural Philosophy* – *Principia* for short) has been called the greatest single achievement of the human intellect. It is certainly difficult to think of a rival. In the writing of it, Newton had come to the verge of exhaustion. His absent-mindedness became the subject of many stories. For example, his friend Dr Stukely called, and became so hungry waiting for Newton to appear that he removed the cover from a plate, and ate the cold chicken underneath. When Newton finally came in, he lifted the cover, muttered in a puzzled voice, 'Dear me. I thought I had not dined,' and replaced it. But the sheer strain involved in the labour was no joke. It left Newton indifferent to science, and eventually led to a nervous breakdown. Newton was forty-four at the time of publication, and still had forty years of life ahead of him. But his original scientific work was at an end.

With the *Principia*, the science of astronomy comes of age. Kepler had worked out the general laws of the motion of the planets; but Newton's mathematical apparatus was so powerful that he was able to account even for the slightest aberrations. He showed, for example, precisely why the earth's axis changes its angle over thousands of years – the 'precession of the equinoxes'; because the earth bulges at the equator, and the sun's gravity pulls it sideways. (The axis, as we know, is tilted; if it was upright, the aberration would not occur.) Through his third law of motion – about the action of bodies on one another – he was able to account for every slight variation in the orbits of the planets. (A minor exception was the perihelion of Mercury – the fact that Mercury's orbit slowly changes its position around the sun; this had to wait for the arrival of Einstein for a solution.) And he was able finally to solve the problem that had baffled Galileo – the motion of the tides.

The *Principia* made Newton the most famous man of science in Europe. Yet it would be untrue to say that this was because it was understood. A French reviewer, after praising his obvious genius, went on to object that he had still not *proved* that the motion of the tides was caused by the moon – a remark that reveals a failure to grasp the principal idea of the book. Neither was the fame an unmixed blessing; he had become a household name – rather like Einstein or Freud in the twentieth century – but many people resented his eminence, and attacked him as a crank with an unproven theory; satires and mocking cartoons proliferated. The result was to confirm Newton's lifelong conviction that a thinker would do better to keep his ideas to himself. In fact, during the next decade, he was to drift close to paranoia and insanity.

Fortunately, fate seemed determined to remove him from the dangerous solitude of his rooms in Cambridge. During the writing of the *Principia*, he had been chosen as a member of an eight-man team to protest about the King's interference in University affairs; the infamous Judge Jeffreys overrode their arguments and sent them packing. But in 1688, James II had to flee to France when William of Orange landed at Torbay; Jeffreys was arrested as he tried to escape in disguise, and died in the Tower. And Newton was chosen to represent the University as a Member of Parliament. He began to spend some time in London, and found he liked it. He met the philosopher John Locke, and the two spent hours discussing theology, particularly the nature of the Trinity. As a result, Newton was bitten by the theology bug, and turned his immense intellectual powers to the problems of biblical chronology. For the remainder of his life, he divided his time between the Bible and alchemy, pouring more energy and enthusiasm into these subjects than he ever had into mathematics.

In 1696, Newton's old friend, and ex-student, Charles Montagu, the Chancellor of the Exchequer, obtained for him the post of Warden of the Mint. After three years he was appointed Master of the Mint, a position which he held until his death. There were few excursions into science or mathematics, yet these few revealed that his intellect was as sharp as ever. In 1696, the mathematician Jacob Bernoulli challenged the mathematical world to solve two problems about curves. He allowed six months for their solution; at the end of that period, only Leibniz had succeeded in solving one of them, so Bernoulli extended the period for another year. Newton received the problems one day when he came home, exhausted, from the Mint. He solved them both that evening, and sent the solutions to the Royal Society the next day. Although they were anonymous, Bernoulli is said to have exclaimed when he saw them, 'Ah, I recognize the lion's paw.' Many years later, Leibniz proposed an equally difficult problem about the equation of a curve. Once again, Newton solved it in an evening.

In 1704, the long delayed publication of his second major work, *Opticks*, confirmed his scientific genius, and reminded Queen Anne that Britain's greatest scientist was still a commoner; she remedied this by knighting him in the following year. In 1711, Newton finally allowed his treatise on the calculus, *De Analysi*, to be printed – he had given Barrow a handwritten copy more than forty years earlier. In the following year, he became involved in the absurd and bitter controversy with Leibniz about who had discovered the calculus first; this episode exhibits some of the worst aspects of Newton's character, but need not detain us here. Although he had now given up serious scientific – or mathematical – work, he still worked incessantly on problems connected with the Bible; his *Chronology*, which now strikes us as nonsense, was as famous in its time as the *Principia*. And he continued working on these chronological questions throughout his sixties and seventies – the unpublished material that he left behind would fill fifty large volumes. Today we regard his alchemical and theological speculations as aberrations of genius – like Kepler's *Harmony of the World*. But it is worth making the imaginative effort to grasp that both Newton and Kepler regarded these works as the crowning achievements of a lifetime devoted to 'justifying the ways of God to man'. Both of them would have been shocked by the modern view that science is merely an attempt to understand the physical universe.

Newton died in 1727, at the age of eighty-four, and was buried in Westminster Abbey. His work was the culmination of two thousand years of astronomy. It was also the beginning of the most exciting epoch so far.

Illustrations from a French educational aid demonstrating the principle of gravity

CHAPTER SIX

THE EXPLORERS

In May 1661, King Charles II of England – newly restored to the throne after the death of Cromwell – peered through a powerful telescope and saw the moons of Jupiter, and the ring around Saturn. He was so impressed that he agreed to grant a charter to a Royal Society for Improving Natural Knowledge. The French king, Louis XIV, not to be outdone, announced the formation of a Paris Academy of Science in 1666. And, since the Sun King had more money than the indigent King Charles, he sent his envoys to the ends of Europe to attract eminent scientists. These included the distinguished Dutchman Christian Huygens, who had first discovered the rings of Saturn (which Galileo had taken for satellites), the Italian astronomer Giovanni Domenico Cassini, and a brilliant young Dane named Ole Roemer. Huygens soon justified his appointment by inventing the pendulum clock, while Cassini discovered that Saturn has a double ring – the gap between the two is still called Cassini's Division.

As director of the Paris Observatory, Cassini was one of the first astronomers to have fairly generous sums at his disposal. So in 1671 he was able to despatch a young assistant named Jean Richer to the colony of Cayenne in South America (French Guiana) to study the way light behaved in the tropics. And when Richer was already in Cayenne, it struck Cassini that here was an excellent opportunity to try and discover the distance of Mars. The method involved the use of parallax which we have discussed in connection with Tycho Brahe – that is, the fact that an object appears to change position when viewed from different places. The distance of about six thousand miles between Paris and South America made an ideal 'base line' for the necessary triangle. Richer made careful observations of the

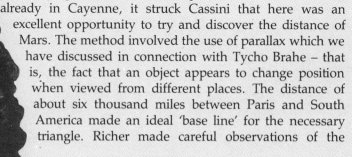

Christian Huygens

Donati's comet seen at Paris, 4 October 1858

position of certain bright stars close to Mars, and Cassini made similar observations in Paris. When the calculation was finally worked out, the result seemed staggering; Mars was apparently about forty million miles from the earth. It was one of the first intimations of the true size of the solar system.

But this result had an even greater significance. Kepler's Third Law stated that the cube of a planet's distance from the sun is the square of its period of revolution. Kepler had also worked out the relative distances of the planets from the sun. That information was useless until someone could establish a planet's actual distance. But as soon as Cassini had this figure, he could work out the distance from the earth to the sun. He reached the conclusion that this was eighty-seven million miles – only about six million miles short of the best modern measurements.

Richer made another interesting observation in French Guiana – that a pendulum beat more slowly there than in Paris. When this observation came to the attention of Isaac Newton, he was able to supply a reason. Cayenne is closer to the equator than Paris. And if the earth has a bulge at the equator – caused by its motion on its axis – then Cayenne would actually be further from the centre of the earth than Paris, so gravity would be weaker there. In fact, Cayenne *is* thirteen miles further from the earth's core than Paris. This distance sounds small enough until we realize that it is equivalent to going up thirteen miles in a balloon. . . .

Cassini was, unfortunately, a thoroughly unpleasant man, opinionated, mean and jealous. When Richer came back from Cayenne in 1673 and found himself famous in Paris, Cassini had him sent off to the provinces to work on military fortifications – one of the standard chores of mathematicians over the centuries. He sank into obscurity and was forgotten.

The next major step was taken by the young Dane Ole Roemer, who arrived in Paris in the year Cassini worked out the distance of Mars. He apparently found the observatory disorganized and uncomfortable – one astronomer was even sleeping on a window-sill. Roemer was given the task of studying the moons of Jupiter. Cassini had already done some brilliant observational work on these moons, and had published an ephemeris – a table of their positions.

Looking at this table, Roemer observed that the periods of their revolutions seemed to be irregular; that is, the moons were sometimes late and sometimes early in making their appearance. They were late, Roemer noticed, when Jupiter was furthest from the earth, and early when it was close. One autumn night in 1674, Roemer was observing Io, the innermost moon, and noted that it was three minutes late. Suddenly, Roemer had the inspiration that was to make his name immortal. Suppose it was the *light* from Io that was three minutes late, not the satellite itself? After all, Cassini had worked out that the sun was eighty-seven million miles away. That meant that when the earth was furthest from Jupiter – round the other side of its orbit – the light had an extra 174 million miles to travel. Surely that ought to take it a little longer?

With the aid of Cassini's ephemeris of Jupiter, and his knowledge of the distance from the earth to the sun (known as the Astronomical Unit – because it is a more convenient way of measuring vast distances than in miles), he could work out an approximate speed of light: 141 000 miles per second. Again, he was out by 24 per cent – the modern determination is 186 000 – but it was a remarkably accurate estimate for the time.

Predictably, the self-important Cassini was upset by this theory of his new assistant. Perhaps he was resentful at not noticing the time discrepancy himself. He argued that it *was* Io that was late, and that some irregularity in the orbit must account for it. But Christian Huygens instantly saw that Roemer must be right. So did Isaac Newton in England; his experiments nine years earlier had convinced him that light consists of particles. And a particle must travel at a finite speed. As for Huygens, he continued to think about the subject of light, and in 1690 produced the theory that finally replaced Newton's – that light is a form of wave. The actual experimental proof was provided by Thomas Young in 1803.

Roemer, like Richer, discovered that Cassini was not fond of subordinates with minds of their own. But he was undismayed by the rejection of his theory of the speed of light; in 1681 he returned to Copenhagen, where he had a distinguished career as an astronomer, and even became burgomaster – an office equivalent to chief of police. He died in 1710 at the age of sixty-six. In 1725, a British astronomer named James Bradley found a more accurate method of determining the speed of light, and recalled the now forgotten theory of the Dane. The detestable Cassini died in the same year, and it is doubtful that he heard that Roemer had been vindicated. Probably it would

A nocturnal, or timepiece using the pole star and the two Bear constellations
Overleaf: The Royal Observatory in Greenwich Park, 1680

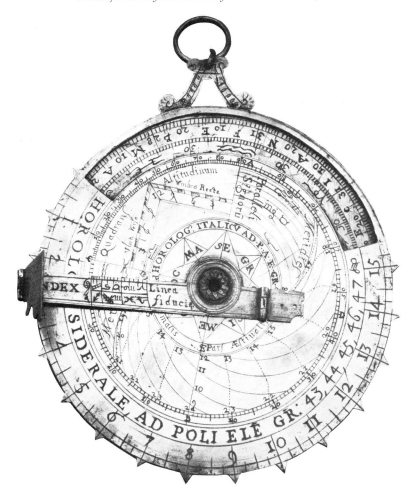

have made no difference; Cassini was one of the last of the major astronomers to reject Copernicus's theory. . . .

But the story now has to shift back to England.

It was not entirely the love of philosophy that prompted Charles II to grant a charter to the Royal Society. Like Rudolph of Prague, he needed star tables to guide his mariners to distant ports. A proposal by a Frenchman, Le Sieur de Saint-Pierre, to use the moon to determine longitude had been rejected, because the moon's orbit was too complex. (This was in 1674, before Newton applied his intellect to the problem.) What was necessary – said the committee set up to investigate this problem – were star maps and a moon-table extending over a nineteen-year period.* (We may recall that figure in connection with the great Stonehenge computer.) So the King heaved a sigh – for he lived off the charity of Louis XIV – and appointed an Astronomer Royal, a young man named John Flamsteed. He also decided to build the Royal Observatory in a park at a place called Greenwich, where there was a convenient hill. But that was as far as he

The first observation of Venus in transit across the face of the sun, 1639

could go. He ordered that the £500 to build the observatory should be raised by selling old gunpowder for blasting, and cut the cost further by providing old wood from a demolished gatehouse in the Tower and old brick from a fort at Tilbury. So the great Greenwich Observatory was built on a shoestring. There was not even money to acquire instruments – the King told Flamsteed to build his own. On a salary of £100 a year, Flamsteed's enterprise was definitely limited. Yet over the next twenty-five years or so, he collected the largest and most accurate body of observations ever made. This, in due course, brought him into violent conflict with Sir Isaac Newton, then President of the Royal Society; for Newton wanted Flamsteed's lunar observations to check his theory of gravity, and Flamsteed seemed to regard them as his own property, and refused to give them up. This was another of the controversies that brought bitterness to Newton's later years.

Flamsteed was helped, in the early days, by a young and enthusiastic amateur named Edmund Halley, son of a wealthy soap-maker. The twenty-year-old Halley was supposed to be taking a degree at Oxford; but at the first suggestion that he might be able to make some independent observations, he decided to leave. On 7 November

* The period, which is actually just over eighteen years, is called a saros.

1677, Mercury was due to cross between the earth and the sun – an event known as a 'transit'. A suitable place to observe this was the island of St Helena, off the west coast of Africa (where Napoleon would one day spend his exile). But Charles II did not have the money to finance the trip. Halley persuaded the East India Company to take him there free, and he spent a year on St Helena, mapping the stars of the southern hemisphere and making accurate observations of the transit. As he did so, it struck him that this could be the ideal way of determining the Astronomical Unit – the exact distance between earth and sun; not by observing a transit of Mercury, which was too close to the sun to give a good degree of parallax, but of Venus – the planet midway between Mercury and earth. Unfortunately, no transit of Venus was due until 1761 – by which time Halley knew he would be dead. (He was born in 1656.) But Halley nevertheless alerted the world of astronomy to the coming transit; and in due course, astronomers from all over the world launched the biggest joint operation ever to observe it.

When Halley returned from St Helena, at the age of twenty-two, he was a celebrity, and the King immediately made him a Fellow of the Royal Society, and ordered Oxford to grant him a degree without examination. Halley went on to become one of the finest astronomers of his time – eventually he succeeded Flamsteed as Astronomer Royal. Four years after his return to England, there occurred the event with which his name is indissolubly linked. In August 1682, Flamsteed observed a comet from the Greenwich Observatory. That was nothing remarkable; there are usually several comets every year, most of them invisible – or only just visible – to the naked eye. But this comet quickly became the most brilliant object in the sky, with its white tail streaming against the stars like some immense stationary sky-rocket.

Ever since ancient times, comets had been regarded as harbingers of disaster. The Romans called them 'hairy stars' because their tails looked like hair (coma is Latin for hair, or a ray of light), and the Emperor Vespasian remarked of the great comet of AD 79: 'It doesn't scare me – I'm bald. Let the King of the Parthians worry about it – he's hairy.'[1] (His confidence was misplaced. He died in the same year.) There had been a great comet in 1066, at the time of the Norman Conquest – it is portrayed in the Bayeux Tapestry. A comet that appeared in 1456 is said to have worried the Turks who were besieging Belgrade – and who were then defeated by János Hunyadi. Even Tycho Brahe felt obliged to add a chapter to his book on the comet of 1577, explaining that it probably signified war, disease and religious upheavals. The sixteenth-century 'magician' Paracelsus was one of the first to raise his voice against this superstition, in a book denouncing all the 'catastrophe theories'. But by the time of Halley, there was still a widespread belief that comets were heavenly messengers, sent to foretell disaster.

Halley had one advantage over former astronomers: as a friend of Isaac Newton, he knew that every body in the heavens is under the influence of gravity. Which meant that as a comet approached the sun – as they all seemed to – it would have to do a U-turn in the heavens. Halley began studying every account of comets he could find. He discovered observations on a bright comet in 1607, just seventy-five years earlier, and on another in 1531, seventy-six years before that. Then there was another bright comet in 1456, and another in 1380. The figures seventy-five and seventy-six seemed to recur as the intervals between them. Could they all be the same comet? If so, then presumably it must be a member of our solar system, like the planets, but with a period of revolution of about three-quarters of a century.

Kepler – who had written a report 'On the hairy star that appeared in 1607' – had

decided that comets travel in a straight line, 'from infinity to infinity'. But if Halley was correct, they must travel in an enormous ellipse, just like the planets; moreover, the comet Kepler had seen in 1607 must have been the comet that caused such a sensation in the night sky in 1682. And if that was so, then its next appearance should be in 1758, by which time Halley himself would be a hundred and two.

Since Halley died at the age of eighty-six, he never knew whether his prophecy was correct. But when his comet was observed on Christmas Day, 1758, by a German astronomer named Johann Palitzsch, it was clear that he was right; the comet was named after Halley, and has continued to appear at intervals ever since. (Its next appearance is due in 1986.)

Other astronomers searched the records for earlier appearances of Halley's Comet, and discovered that astronomers had seen it as long ago as 1058 BC. It had even appeared in the year 12 BC, which has led to the interesting suggestion that it might have been the 'star of Bethlehem' seen by the wise men, announcing the birth of Jesus; but although historians have questioned the accuracy of the birth date usually attributed to Jesus, it seems reasonably certain that 12 BC is too early. (The likeliest date is between 7 and 5 BC.)

But what are comets? It is chastening to have to record that we are still not certain. One theory suggests that they are made up of interstellar dust, collected by the sun as it passes through the cloud. In 1930, the Estonian astronomer Ernest Öpik suggested that there must be some vast cloud out in space, perhaps as much as a light-year away (the distance light takes to travel one year – nearly six million million miles). The Dutch astronomer Jan H. Oort developed this theory in 1950, suggesting that the 'cloud' (known now as 'Oort's Cloud') is made up of fragments left over after the solar system had condensed from a vast cloud of whirling gas. Oort's Cloud would be made up of freezing gas. Periodically, some disturbance – like a passing star – would dislodge chunks of it from its orbit, and send them falling in towards the sun. So a comet would be a kind of dirty snowball, a few dozen miles in diameter. As it approaches the sun, the heat makes the freezing gases evaporate and they stream behind the comet's head in a glowing tail, pushed back by the solar wind. The tail vanishes once the comet recedes into outer space. It seems fairly certain that comets have a low density, as well as a fairly small size (a few dozen miles across); a comet passing between the moons of Jupiter has failed to cause any obvious disturbance in their orbit. This also means that, unlike the planets, the life of a comet is relatively short – a matter of thousands rather than millions of years. A comet observed by an Austrian captain, Wilhelm von Biela, in 1826 reappeared – on schedule – in 1832 and 1839; but when it appeared in 1846, it had split into two. In 1852, the twin halves had separated even further, and by 1866, it had simply ceased to appear. An astronomer named Weiss calculated that the earth would pass through the comet's orbit in 1872; when this happened, there was a spectacular display of 'shooting stars' – in Italy, over thirty-three thousand were counted in six and a half hours. It seems, then, that comets are gradually torn apart by the sun – and possibly by other heavenly bodies – and disintegrate into clouds of icy fragments. So a catastrophe like that described in H. G. Wells's story *The Star* – in which a passing comet causes earthquakes and other immense upheavals – is, to put it mildly, unlikely.

Although Halley is best known as the man who first understood comets, astrono-

Opposite: (above) Harold is told of the comet. Bayeux tapestry; (below right) Photograph of Halley's comet taken in 1910; (below left) Edmund Halley, Astronomer Royal

mers remember him for a far more important achievement: the discovery that the stars also move. His studies of ancient reports of comets also led Halley to examine the star tables of ancient astronomers like Ptolemy. He was intrigued to note that the brightest stars – like Sirius, Arcturus and Aldebaran – were no longer in the precise positions given by the Greeks. He might well have dismissed this discrepancy as resulting from the inaccuracy of their observations. But he saw another possible explanation: could they be brighter because they were closer than the other stars? If so, their movement could, of course, be caused by the parallax effect – meaning that it was the earth and sun that had moved, not the stars. Or it could mean, quite simply, that the stars themselves moved. It was a breathtaking assumption – the phrase 'fixed stars' had become part of the language. And it had an interesting corollary. If the movement was genuine, then the stars must be an almost inconceivable distance from the earth – otherwise the movement would not be so small in the fifteen hundred years since Ptolemy. The universe of the astronomers was becoming terrifyingly vast. . . .

Before we leave Halley, it is worth adding an intriguing footnote. The body of Jewish law and legend known as the *Babylonian Talmud*, compiled between the third and fifth centuries AD, tells a story of two rabbis on a long sea journey. One of them has brought flour with him, and explains that he felt this necessary because 'there is a very bright star that appears every seventy years and deceives sailors'. A French scholar, M. Renaudot, calculated that the sages' voyage took place about AD 66, and the rabbi was probably referring to Halley's Comet, which appeared in that year. Whether or not the comet referred to is Halley's, it seems clear that the authors of the *Babylonian Talmud* knew something that was rediscovered by Edmund Halley.[2]

Halley died in 1742, his last great achievement being a detailed observation of the moon for the eighteen-year period between eclipses – a saros. This work was so accurate that it finally made it possible to determine a ship's longitude at sea by the position of the moon – the notion that had led Charles II to appoint his first Astronomer Royal in 1675.

Halley's successor as Astronomer Royal was James Bradley, the man who was finally to confirm Roemer's belief that light had a definite speed. The story of how this came about is as strange as anything in the history of astronomy. In 1725, Bradley (who was not yet Astronomer Royal) became involved in an attempt to try and discover the parallax of the stars – whether they appeared to move as the earth revolved around the sun. He set up a telescope that pointed directly overhead, reasoning – correctly – that if the star's light had less atmosphere to penetrate, it would be refracted less. After a mere fourteen days, there appeared to be a definite shift in the position of the star Gamma Draconis. This was obviously too good to be true; the star couldn't be *that* close. As the observations continued, his bewilderment increased. If the movement *was* parallax, then it should be greatest in December and June. But this 'shift' was at its smallest in these months, and at its greatest in March and September.

Bradley tried every possible explanation – unknown effects of atmosphere, unknown movements of the earth. . . . None seemed to explain his observations. The breakthrough came one day when he was travelling on a Thames steamer, and noticed that the pennant had changed direction. It was not, of course, a change in the wind, but in the direction of the boat. Could that analogy explain what was happening to the stars? Perhaps we were seeing a similar effect as the earth travels about the sun.

How could the earth's speed explain his odd results? Then the answer came. If you are looking at a star directly overhead, and your telescope is moving along at about eighteen miles per second (because the earth is also moving), what would happen to the beam of light from the star? Let us suppose it enters the middle of the telescope lens. Then it has to pass down a 24-foot tube *which is in motion*. By the time it reaches the bottom of that tube, the tube will have moved along slightly, and the light will no longer emerge from the centre of the lens, but from slightly to one side. So you will see the star slightly ahead of its actual position. *Very* slightly, it is true. But then, the effect Bradley observed *was* very slight.

Most writers on this phenomenon rely on the same simple analogy. Suppose rain is falling from directly overhead while you are driving your car. The rain hits your windscreen as if it is *slanting* down. The faster you go, the more it seems to slant; and if

Astronomical chart showing the constellations Corona Borealis, Hercules and Boötes, 1750

you were travelling as fast as it was falling, it would seem to be coming from directly ahead, parallel to the ground.

So Roemer was vindicated; light *must* take time to travel.

From then on, it was simply a matter of refining the apparatus to measure the exact speed of light. One of the major points of contention was whether light was made up of waves, as Huygens suggested, or particles, as Newton believed. This question was apparently settled in 1800 by an English scientist, Thomas Young, who showed that if light is passed through two slits very close together, some of the light from one slit 'cancels out' some of the light from the other, producing a dark band. Flying particles cannot cancel one another out, but waves can. (Quantum theory was later to decide that light behaves like both waves *and* particles, and that probably neither concept corresponds to the reality.) Fifty years later, Jean Foucault performed a classic experiment showing that light travelled slower in water than in air, as Huygens had also predicted. (For some odd reason, Newton thought that refraction – the way things seem to bend in water – was caused by the increased speed of light particles in water.) The stage was set for the appearance of Einstein and Max Planck. . . .

The most extraordinary thing about Bradley's discovery of the displacement or aberration of light is his incredible accuracy. In his paper describing his discovery, he mentioned that the aberration amounted to no more than two seconds of an arc, and added that he could have detected it even if it had only been one second. (A second of an arc is one-sixtieth of one-sixtieth of a degree – or the apparent size of a pinhead at a distance of 225 yards.) And even with apparatus of this accuracy, no parallax could be detected in the stars – which, as Bradley pointed out, suggested that they were an unimaginable distance away.

Bradley was one of a new breed of astronomers who recognized that new advances depended on a degree of accuracy that even Kepler or Galileo would have regarded as impossible. Halley himself, that enthusiastic and lovable man, had been oddly slap-dash about the finer points of observation – so his eighteen-year moon-tables were far less useful than they should have been. But Bradley's discovery that stars had no visible parallax made everyone aware that the future of astronomy lay in unprecedented accuracy combined with superb instruments. The day of the enthusiastic amateur seemed to be drawing to a close.

Yet the next great British astronomer was conclusively to disprove this. When Friedrich Wilhelm (later Sir William) Herschel was born in Hanover in 1738, the Elector of Hanover was King George II of England. His father was a musician in the Hanoverian army, and Herschel himself became for a time an oboist in the Hanover Foot Guards. When the Hanoverian army became involved in the Seven Years' War (1756–63 – a global conflict that could be called 'World War Minus One') Herschel – whose inclinations were more scholarly than martial – came to England, and soon became a highly successful organist and music teacher. He settled in Bath, that fashionable watering place, and soon found himself prosperous enough to gratify his passion for study. He taught himself Latin and Italian, and then turned to science. At this point, at the age of thirty, he discovered Newton's *Opticks* – an excellently written book, far less dense than the *Principia* – and overnight became a devoted astronomer. At first he hired a small telescope; but it proved unsatisfactory. A good telescope was beyond the means even of a prosperous organist, so he decided to teach himself to grind his own mirrors. It proved to be heartbreaking work; attempt after attempt was a failure. But Herschel possessed German thoroughness; he ground on – literally – until, at his two-hundredth attempt, he made himself a good 5-inch mirror.

His astronomical work had become so absorbing that he returned to Germany, and brought back his sister Caroline to keep house for him. She proved to be a born lens-grinder – and astronomer. Together, they made better telescopes than any then in existence.

In 1781, when Herschel was forty-two years old, his persistence paid off in a most spectacular manner. On 13 March, the music master was peering at the stars in the constellation of Gemini when he observed one that seemed brighter than the others. As he was hoping to detect the parallax of the stars, he naturally paid close attention to his new discovery. A more powerful reflector – $6\frac{1}{2}$ inches – magnified his 'star'; this convinced him that it could not be a star after all, since the stars are too far away to be greatly magnified by more powerful lenses. It had to be a comet. A comet in the band of the zodiac – that strip across the sky in which the sun and planets move. Yet even Herschel, with his inborn optimism, could not dare to hope that he had discovered a planet. Besides, everybody knew that there were only seven planets in the solar system, seven being a magic number. The next night, Herschel looked at it again; it *had*

Ancient astronomical instruments at the observatory of Jaipur in Rajastan, India

moved. So it had to be a comet. In April, he reported the comet to the Royal Society; the Astronomer Royal, Nevil Maskelyne, soon concluded that the new object was more like a planet than a comet. There was only one way to find out – to track its orbit. It was soon clear that this was circular. (Newton's laws enabled mathematicians to decide this far more quickly than Kepler would have been able to.)

The unknown organist had discovered the first new planet in historical times. England was prompt to recognize him. He was elected to the Royal Society in the same year; and the need to make a living by playing the organ was finally behind him. Herschel became King George III's private astronomer at a salary of £200 a year. In 1788, he married a wealthy widow, and could at last afford to buy the best equipment.

It is an interesting thought that if Uranus had been elsewhere in the heavens on that night of 1781, Herschel might well have remained an obscure organist and devoted amateur astronomer, instead of becoming the most famous astronomer of his time.

For his attempts to observe the parallax of stars were entirely unsuccessful. As it was, he became a celebrity overnight, and a favourite of the King – whom he tried to repay by naming his new planet 'Georgium Sidus' – George's Star. This was as unsuccessful as Galileo's attempt to call the satellites of Jupiter after the Medici family. Some astronomers called it Herschel (a name that persists in some astrological ephemerides). But the classical tradition finally prevailed, and by the mid-nineteenth century, most astronomers had accepted a suggestion by the German Johann Elert Bode, and called it Uranus.

It was appropriate that Bode should have the distinction of naming the new planet, for he had predicted its existence before it was discovered. It was Bode who redis-covered a mathematical curiosity, first observed by Titius of Wittenberg in 1766, and made it famous. Take the numbers 0, 3, 6, 12, 24, 48, 96 and 192 – all (except 0) double the preceding number – and add 4 to each of them, making 4, 7, 10, 16, 28, 52, 100 and 196. These numbers correspond fairly accurately to the actual distances of the planets from the sun, if earth (the third planet) is associated with the third number, 10. One-tenth of the distance between earth and the sun is about nine million miles; and Mercury *is* four times nine million miles from the sun; Venus is seven times, Mars is sixteen times, and so on. But when Bode came across this 'law' in 1772, there was no planet corresponding to number 28 in the vast gap between Mars and Jupiter. And, of course, Uranus was not known. In due course – nine years later – Uranus turned up as number 196, just where Bode's Law predicted – or near enough to make an incredible coincidence. Unfortunately, as we shall see, the discovery of the next planet, Nep-tune, made nonsense of the 'law' – being at a distance of 300 instead of 388 units. Oddly enough, the most recently discovered planet, Pluto, *does* fit the figure that should be reserved for Neptune. . . .

In spite of the discovery of Uranus – which could have happened to anyone – Herschel's greatest achievements were in connection with the stars. He was the first man to grasp the picture of the universe more or less as we do today, with its unimaginable distances. His most amazing achievement was to realize that *we* are a part of the galaxy called the Milky Way, and that the 'band' of stars across the heavens we call the Milky Way is simply a wheel-shaped cluster *seen sideways*. It was Herschel who confirmed Halley's theory that the stars also move; he even specified that our own sun is moving in the direction of the constellation Hercules. He suggested that our own solar system is somewhere towards the centre of our galaxy, and here he was mistaken – it is closer to the edge. Yet the totality of his insights was astonishing, and over his long lifetime (he died at the age of eighty-four) he probably did more for astronomy than any man except Kepler. He also built the largest telescope ever constructed – 40 feet long, with a 48-inch mirror – and the first time he used it, in 1787, he discovered that Uranus had satellites (Oberon and Titania); in 1789 he also found two new satellites of Saturn. Herschel was knighted in 1816, at the age of seventy-eight; he died in 1822.

His son, John Frederick William (later Sir John) Herschel (1792–1871), continued his father's work, mapping the stars of the southern hemisphere, and cataloguing binary stars (stars that circle round one another, like Sirius A and B) and nebulae. In August 1835, his name became known to the wider public in an unexpected manner; a New York newspaper called *The Sun* printed a series of articles, which it claimed to be reprinted from the *Edinburgh Journal of Science* (a non-existent magazine); in these, Sir John Herschel reported how, through an enormous telescope at the Cape of Good

Hope, he had seen living creatures on the moon – furry, winged men resembling bats. The circulation of *The Sun* climbed overnight to twenty thousand – enormous for those days – and Edgar Allan Poe, who was working on the second part of a 'moon hoax' story called *The Unparalleled Adventures of One Hans Pfaal*, decided to abandon it, feeling he had been outdone. But nine years later, Poe managed to score a similar success with his own 'Balloon Hoax', claiming that nine men had crossed the Atlantic in seventy-five hours in a balloon and landed in South Carolina. Both hoaxes took in most of the American public for a matter of weeks.

William Herschel's 40-foot reflecting telescope mounted on a revolving platform

Still, it would hardly be fair to blame the public for gullibility. Advances in science were now progressing at such a pace that almost anything was possible. Newton's *Principia* had opened up new continents of knowledge for exploration. In 1774 the Astronomer Royal, Nevil Maskelyne, even succeeded in weighing the earth – a feat that certainly *sounds* like a hoax. He did this with the aid of an ordinary plumb-line, and Mount Schiehallion in Scotland. On either side of the mountain, he observed how far the plumb-line deviated from the vertical – attracted by the gravitational pull of the mountain. Then he worked out the weight of the mountain (which was solid granite) and compared the pull of the mountain with the pull of the earth. He estimated that the average density of the earth is four and a half times that of water. (We make it nearer five and a half.) Its weight was 6592 million million million tons.

The man who continued the work Newton began in the *Principia* – and very nearly brought it to completion – was the French astronomer and mathematician Pierre Simon de Laplace, about whom Poe spoke so slightingly in connection with his own *Eureka*. Laplace was born in Calvados in 1749, the son of a farmer. At eighteen, he went to Paris to seek his fortune, and sent a note to the encyclopedist and astronomer Jean Le Rond d'Alembert, who ignored it. Laplace then went back to his lodgings and wrote d'Alembert a brilliant letter about mechanics. This succeeded where his previous letter of recommendation had failed; d'Alembert became his patron, and secured his appointment as professor of mathematics at the Paris Military Academy. Laplace's biography is the story of a man determined to 'get on' in the world, who eventually became a marquis. But his major work was the vast *Traité de Mécanique Céleste* (or *Celestial Mechanics*), whose purpose was to work out the extent to which each member of our solar system affects all the others. It is, obviously, an extraordinarily complex problem, since every body affects every other body, and all are moving. It took him twenty-six years, and five large volumes, to accomplish it. In a footnote to one volume, he suggested casually that the solar system could have had its genesis in a great ball of gas which had slowly cooled down, forming the sun and planets. This 'nebular hypothesis' had been anticipated, incredibly enough, by the German philosopher Immanuel Kant as long ago as 1755; Kant had also suggested that the Milky Way was a vast, lens-shaped cloud of stars – anticipating Herschel by half a century – and that other similar 'island universes' existed in space. (Considering the state of astronomical knowledge at the time, this seems like sheer inspiration.) Laplace was apparently unaware of Kant's theory, and his *Celestial Mechanics* ended with a triumphant demonstration that our solar system is absolutely stable, and can be expected to last – more or less – forever. If he had read Kant's *General History of Nature and Theory of the Heavens*, he might have also noted an important suggestion that the earth's rotation is being slowed down by the movement of the tides. (This has since been verified.) So he confirmed the basically Newtonian view – that fitted in so well with the outlook of his time – that the universe was a gigantic piece of clockwork, which had been set in motion by God, and could expected to run forever. . . .

What was slowly becoming clear was that the various sciences could not be kept in watertight compartments; they were interconnected. In 1760, five years after the great Lisbon earthquake, a geologist (and clergyman) named John Mitchell had suggested

Sir John Herschel continued his father's work on binary stars and nebulae.
Portrait by Julia Margaret Cameron

that earthquakes and volcanoes are connected, and that our earth may be a great boiling cauldron beneath the surface. He was, in effect, supporting Kant's nebular hypothesis (which appeared in the same year as the Lisbon earthquake) – that is, suggesting that the stars and planets were not created by God on the first day of creation, but that they may have evolved out of fire and chaos over vast periods of time. In 1795, a Scottish farmer and geologist, James Hutton, published a *Theory of the Earth* in which he showed that the stratification of rocks and the formation of fossils was still going on. Then there was the mineral surveyor and civil engineer, William Smith, who crawled around in mines, and recorded the different geological layers of the earth, each with its characteristic fossils. Yet when, in 1830, Charles Lyell published the first volume of *Principles of Geology*, demonstrating conclusively that the earth must be millions of years old, the revelation came as a shock to the European intellectual world; men instinctively clung to the old theory of the seven days of creation, with the earth as the focus of the story.

This increasing knowledge of the heavens and the history of our earth was accompanied by a slowly expanding insight into the secrets of matter itself. In 1752, the American scientist and philosopher, Benjamin Franklin, sent a kite aloft during a thunderstorm and received an electric shock from a storm-cloud. At that time, there was great concern about the danger of lightning igniting stores of gunpowder. (For example, in 1769 the explosion of an arsenal at Brescia, in northern Italy, blew up most of the town.) Franklin's kite experiment revealed the solution to this problem: lightning-conductors. At the time of the Brescia explosion, Franklin, who had been appointed Pennsylvania's agent in London, happened to meet a Nonconformist clergyman named Joseph Priestley, who became so fascinated by Franklin's talk about electricity that he decided to begin his own experiments.

But Priestley's real interest was in chemistry. In the brewery next to his home in Leeds, he studied the gas given off by the bubbling vats, and discovered that it would extinguish a lighted taper. Dissolved in water, it made a pleasantly fizzy drink like Seltzer-water. The gas was, of course, carbon dioxide; it had already been discovered by the Scottish chemist Joseph Black. By heating red oxide of mercury, Priestley obtained another gas, which, far from extinguishing a taper, made it burn more brightly. This was oxygen. But Priestley was prevented from grasping the significance of his discoveries by a mistaken theory. Like most of his contemporaries, he believed that when a substance – like wood – burned, it *gave off* its 'combustible' material; this material was called 'phlogiston'. So Priestley called oxygen 'dephlogisticated air', meaning air that does not contain phlogiston. Fortunately, he described his various experiments to the great French chemist Antoine Laurent Lavoisier,* who repeated them, and finally realized that oxygen is a pure gas – an element, not an impure form of atmospheric air. Lavoisier's explanation of what happens when something burns was, in its way, as epoch-making as Copernicus's discovery about the sun. And it led an English chemist, John Dalton, to realize that when 'elements', like hydrogen and oxygen, combine, they do so in exact proportions – proportions that are always whole numbers. Which in turn suggested that each substance must be divisible into 'minimum quantities' called atoms. The next step in astronomy occurred when some-

---

* The same Lavoisier who, in September 1768, was sent by the Academy of Sciences to investigate a 'great stone' that had hurtled out of the sky near the town of Luce. Lavoisier proved himself a prejudiced reporter; convinced that stones never fell out of the sky, he told the Academy that all the witnesses were either mistaken or lying. The reality of meteorites – fragments of disintegrated comets – would only be accepted by the Academy after the physicist Jean Baptiste Biot had investigated a whole shower of them in 1803.

one realized that these same atoms could provide a key to the mystery of the stars.

Oddly enough, Newton had already held that key, without grasping its significance. It was the prism; or, more specifically, the spectrum of colours that it cast. In 1800, William Herschel began to suspect that the spectrum might offer some clues to the mysteries of sunlight. He tried holding a thermometer in the different colours cast by a prism, to see if any were warmer than others. He was startled to find that the thermometer showed most reaction *below* the red end of the spectrum, where there was no visible colour. He concluded, correctly, that sunlight contains an invisible form of light 'below' the red; it was christened 'infra-red'. Only one year later, a German chemist named Johann Wilhelm Ritter was testing the reaction of light on silver chloride, which it turns black (the same reaction that produces photographs), when he discovered that the greatest effect was produced by a 'light' beyond the violet end of the spectrum; this invisible light was christened 'ultra-violet'.

Had Newton produced his spectrum in a different way – by letting light escape from a slit, rather than through a round pinhole in his blinds – he would have seen dark, vertical lines interspersed between the colours. This was noticed in 1802 by an English chemist named William Hyde Wollaston; yet he failed to grasp its significance. This was left to a Bavarian optician and physicist named Josef von Fraunhofer, who was working on the old problem of how to eliminate chromatic aberration in telescope lenses – the problem that had led Newton to invent the reflecting telescope. Fraunhofer not only noticed the dark lines in the spectrum of sunlight: he studied them carefully, and accurately mapped out over five hundred of them. (They are still known as Fraunhofer lines.) Then he placed a prism in a telescope, and examined the spectra of light from the stars. And noticed the dark lines were in different positions.

This should have been one of the most exciting discoveries in nineteenth-century astronomy; in fact, it was – but only when rediscovered half a century later by Gustav Kirchhoff. Fraunhofer had stumbled upon the link between stars and atoms. When any substance is heated, its atoms vibrate, and give off a characteristic light. When this light is passed through a prism, it has its own characteristic Fraunhofer lines. So each element has its own 'fingerprint'. What Fraunhofer had discovered was a way to study what the stars are made of. Then why did he not realize this? Because he never had the means and leisure to devote himself wholly to science. He was one of those men who seem destined to bad luck. The son of a poor glazier, he was the only survivor when the tenement in which he lived collapsed. The instruments he made were the finest that had yet been achieved; yet in class-conscious Germany of the nineteenth century, he was a mere technician. He was allowed to attend scientific meetings, but he was not entitled to address them. Although he finally achieved a degree of recognition – becoming director of the Munich Optical Institute – the glass he ground had got into his lungs, and he died of tuberculosis before reaching the age of forty.

Yet the results of his genius were not entirely wasted. It was his instruments that enabled the Prussian astronomer Friedrich Wilhelm Bessel to realize the dream of the past three centuries, and finally measure the parallax of a star.

Bessel has been called the father of precision astronomy. Like many great astronomers, he began as an enthusiastic amateur. The son of a poor civil servant, he became a clerk in an export firm in 1799, at the age of fifteen. In his spare time he read obsessively, studied languages, and eventually came to astronomy through an interest in navigation. At the age of twenty, he used Halley's observations of his comet to work out a detailed orbit, and sent the paper to Heinrich Olbers, Germany's leading expert on comets. (We shall meet him in the next chapter in connection with the discovery of the asteroids.) Olbers recognized Bessel's talent, and obtained a post at the Lilienthal observatory for his protégé. Faced with the choice of being a prosperous exporter or a poor astronomer, Bessel unhesitatingly chose the latter. For the next thirty years he did invaluable, if pedestrian, work on the cataloguing of stars. He also designed his own heliometer – a device originally intended for measuring the sun's diameter; Fraunhofer made the instrument.

The heliometer is basically a telescope with a 'divided lens' at the far end (this is known as the object glass, as distinguished from the eyepiece). The two halves of the lens can be made to overlap by turning a calibrated screw; this produces a double image of the star. So to measure the diameter of, say, the sun, the astronomer simply turns the screw until the two edges of the 'two suns' are touching, then turns it until the two images coincide; the number of turns of the screw gives him the diameter.

Bessel saw that this could also be used to measure the distance – or apparent distance – between two stars that are close together. And if this distance changes by the slightest amount, then the delicate screw will measure it.

So what Bessel needed was two stars that were apparently close together; one of these two had to be as close to the earth as possible. He decided on a star called 61 Cygni – number 61 in the constellation of the Swan. It was very faint; but it was conveniently close to two other stars, and the speed of its proper motion – its real, independent movement across the sky – suggested that it was closer to the earth than any other known star. In 1837, Bessel focused his heliometer on 61 Cygni, and proceeded to take measurements. Six months later, when the earth was at the far end of its orbit, he did it again. And there *had* been a definite change in the position of the star in relation to the two behind it. It was a very tiny angle – a mere 0.676 seconds of an arc. Yet it gave Bessel the measurement he needed for working out the distance of the star by simple trigonometry, since he knew the length of the base-line of his triangle (the earth's orbit) and the angle made by the apparent motion of the star.

He expected the distance to be great, but the result was breathtaking – thirty-five million million miles. This distance is so immense that it would take light six years to travel it. It later turned out that Bessel had made the angle too large; the correct distance was closer to ten light-years. But this was a small point; the distance of a star had at last been measured. Bessel had accomplished in six months what Herschel had spent most of his adult life trying to do. For astronomy, it was one of the greatest days in its whole history.

Strangely enough, two more men – Friedrich Georg Wilhelm von Struve, a Russian of German origin, and Thomas Henderson, a Scot – also succeeded in measuring the distance to individual stars in the same year. Henderson was able to make his observation on a southern binary star – Alpha Centauri – since he was working at the Cape of Good Hope, while Struve measured the distance to Vega at Dorpat, in Estonia. Sir John Herschel hailed this triple discovery as a triumph over 'the great and hitherto impossible barrier to our excursions'.[3]

It was Bessel who, in 1844, noticed that the star Sirius – the sacred star of the ancient Egyptians – showed tiny movements, which he correctly interpreted as movements around an unseen companion star. (Bessel also discovered another 'eclipsing binary' with an invisible companion, Procyon.) Sirius B, the invisible 'companion', was first

Giant telescope at the Paris Exhibition of 1900

Deuxième année. — N° 8.   Huit pages : CINQ centimes   Dimanche 19 Février 1900

LE PETIT MÉRIDIONAL

ABONNEMENTS   ANNONCES

Supplément Illustré du Dimanche

SIX MOIS | UN AN
France, Algérie, Tunisie. 2 fr. | 3fr.50
Étranger (Union postale). 2 fr. 50 | 5 fr.

Direction, Rédaction, Administration : Rue Henri-Guinier, MONTPELLIER

POUR LA PUBLICITÉ S'ADRESSER
À Montpellier : Rue Henri-Guinier
À Paris : 131, rue Montmartre

Une des merveilles de l'exposition de 1900.
La lunette monstre qui mettra la Lune à quelques lieues de la Terre.

171

seen by the American astronomer (and telescope-maker) Alvan G. Clark in 1862 – by which time Bessel had been dead for sixteen years. It was not until 1914 that another American, Walter Sydney Adams, realized that for Sirius B to be so small *and* so hot, it must be a 'superdense' star – a star in which the atoms had collapsed in on themselves. Such objects came to be called 'white dwarfs'. (We have noted, in Chapter 2, that an African tribe called the Dogon have a tradition that Sirius is a double star with an immensely heavy companion.)

Just as Newton's *Principia* marked the beginning of the great exploratory age in astronomy, so Bessel's measurement of the distance of 61 Cygni marked the beginning of modern astronomy. Had Fraunhofer been heeded this new age might have begun a quarter of a century earlier. It now seems incredible that with 'customers' like Bessel and Struve, Fraunhofer's discovery of the dark lines in the spectrum should have been ignored. Yet in a sense, it was not so strange. Bessel and Struve were great observers –

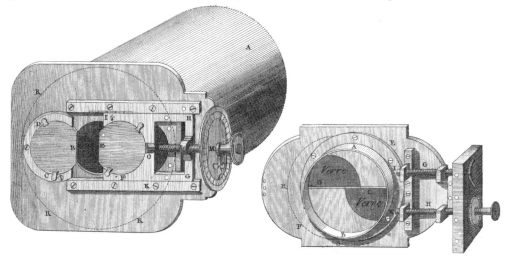

A heliometer, the device used in determining the angular distance between two stars

methodical scientists, not imaginative explorers. The exploitation of Fraunhofer's discovery had to wait for the arrival of another explorer – Gustav Robert Kirchhoff.

This remarkable chapter of intellectual history must begin with the discovery, around 1780, that heated coal produces an inflammable gas; by 1799, the Soho Engineering Works of James Watt – the inventor of the steam engine – was illuminated by two gas jets. When gas lighting finally came to Heidelberg in 1855, it excited a chemist named Robert Wilhelm Bunsen, who grasped its usefulness as a laboratory tool – earlier chemists had been forced to use furnaces. He invented the Bunsen burner in the form we still use today.

The interesting thing about the Bunsen flame was that it could be rendered almost invisible if enough air was allowed into the burner. And this discovery interested Bunsen's friend Gustav Kirchhoff, a professor of physics at Heidelberg University whose lectures were noted for their dullness. For Kirchhoff was interested in the spectra that could be obtained by heating various substances, and passing the light through a prism. An ordinary yellow flame was no good for this purpose, but the Bunsen burner flame was ideal. Bunsen and Kirchhoff were now able to devote themselves to the study of the Fraunhofer lines in the spectra of various elements. The usual method was to dip a platinum wire in the element – sulphur, magnesium or whatever – and hold it in the flame. (Platinum was used because it reacts with so few substances.) The resulting light was passed through a prism and examined.

Kirchhoff made the discovery that should have been made by Fraunhofer half a century earlier: that each element had its own individual Fraunhofer lines – its own fingerprint, so to speak. But these were not the *dark* lines that Fraunhofer observed in sunlight. They were usually exceptionally bright lines. Sodium, for example, had two bright yellow lines in its spectrum.

The simple machine they constructed to examine the various spectra was christened the 'spectroscope'. Bunsen and Kirchhoff invented it in 1859, and in the following year Bunsen had used it to discover two new elements, caesium and rubidium.

Now came Kirchhoff's individual moment of glory. He noted that the bright lines in the sodium spectrum were in the same place as a dark line in the spectrum of sunlight – at the place Fraunhofer had labelled D. He wondered what would happen if he superimposed the two spectra one on top of the other. Surely the two ought to cancel one another out? He tried it, passing sunlight through sodium gas. But the result contradicted his expectation. The two bright yellow lines simply vanished, leaving only a dark line. What had happened?

Then he saw the answer. That particular wavelength of light was being absorbed by the sodium gas. The gas was refusing to let it through, so to speak. So the D line in the solar spectrum proved that there must be sodium gas in the sun's atmosphere. All he had to do was to identify the other dark lines in the sun's spectrum – find out which elements had a bright line in that particular place – and he could state with confidence that that element existed in the sun's atmosphere. He was able to identify hydrogen, iron, magnesium, calcium and several other known elements.

In 1868, when the sun was eclipsed by the moon – so that the only light came from its atmosphere, glowing around the edge of the moon – he was able to observe the same lines in the sun's spectrum – this time, bright emission lines – proving that they came from heated gases in its atmosphere.

In 1835, the materialist philosopher Auguste Comte had made the rash assertion that the composition of the stars is something science will never be able to discover. Kirchhoff's discovery proved him wrong.

An amusing story tells how Kirchhoff's bank manager asked him, 'What use is gold in the sun if it cannot be brought down to earth?' When Kirchhoff was awarded a prize in gold sovereigns for his scientific work, he handed it to his manager with the comment: 'Here is some gold from the sun.'[4]

During the eclipse of 1868, another astronomer, Pierre Jules César Janssen, went to India, taking a spectroscope with him, and observed the lines of a so-far unidentified element in the spectrum. He forwarded his observations to Norman Lockyer – the astronomer who would write about Stonehenge and Egyptian temples – who happened to be the world's leading authority on solar phenomena. Lockyer confirmed that this was an unknown element; it was labelled helium. In 1895, helium was discovered on earth by the chemist William Ramsay.

Only one more major step was needed to transform astronomy into something altogether new and strange. It was taken in 1860 – the year after Bunsen and Kirchhoff invented the spectroscope – by a Scottish physicist named James Clerk Maxwell (1831–79). Three years earlier, Maxwell had made his only direct contribution to astronomy when he proved mathematically that the rings of Saturn could not be solid or liquid, but had to be made up of small solid particles. But his most important theory struck most scientists as impossibly suppositional: that light is only one kind of electromagnetic vibration, and that a whole vast range of undetected vibrations must also exist. We recall that Herschel had discovered infra-red light, and that Ritter discovered ultra-violet radiation. What Maxwell was suggesting was that the same 'scale' could be extended almost indefinitely in both directions, to produce far longer and far shorter waves. He died of cancer at the age of forty-seven only eight years before Heinrich Rudolph Hertz generated radio waves – waves of far greater length than infra-red light. (Housewives will know that steaks can be cooked with radio waves, or 'microwaves'.) Eight years later, in 1895, Wilhelm Conrad Roentgen was observing the luminescence in a cathode ray tube when he noticed that a sheet of paper coated with a barium salt was shining – even though none of the light from the tube was able to reach it. Whatever was causing the paper to shine was *penetrating* a cardboard barrier. Roentgen had discovered X-rays – rays whose wave length is far shorter than that of ultra-violet light, so that it can penetrate many solid substances.

Roentgen's discovery fascinated a French physicist, Antoine Henri Becquerel, who began studying other fluorescent substances hoping to detect X-rays. In the year after the discovery of X-rays, Becquerel placed a uranium compound on a photographic plate; the compound would glow like luminous paint when it had absorbed sunlight. To Becquerel's surprise, his uranium salt still affected a photographic plate when there had been no sunlight for days. He had discovered the phenomenon called radio-activity.

The discovery of radioactivity would have no direct influence on astronomy; but it would open up the world inside the atom – and so enable astronomers and physicists to unite in penetrating the secrets of the origin of the universe. As for radio waves and X-rays, these would also prove to be new and valuable tools for exploring the depths of interstellar space.

But that part of the story would not begin for another half century. . . .

PART III

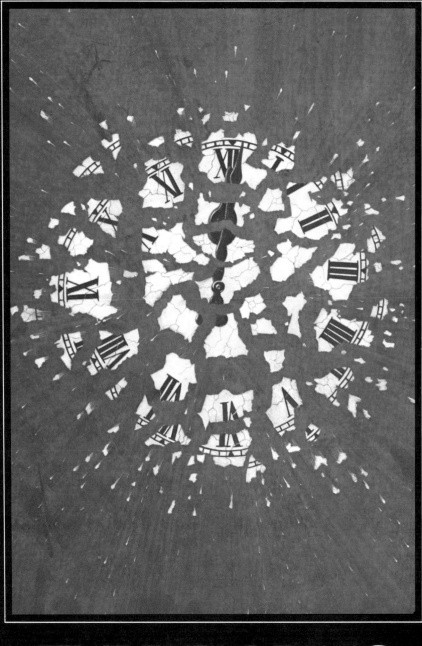

THE EXPLODING UNIVERSE

CHAPTER SEVEN

A VOYAGE ROUND THE PLANETS

I n 1871, James Clerk Maxwell invented an attendant spirit, or demon, whose purpose was to perform imaginary experiments – that is experiments that could not actually be performed in the laboratory. (Scientists called these 'thought experiments'.)* In this chapter, I propose to borrow Maxwell's demon to conduct us on a tour of the solar system.

Without the demon's aid, we would need a rocket and a vast amount of fuel to escape the surface of our planet; in fact, we would need to reach a speed of seven miles a second to overcome the pull of the earth's gravitational field. With the aid of the demon, we can ascend in a transparent shell, at our own speed, making scientific observations on the way.

This pleasant mode of ascent was first achieved in 1783, when a hot-air balloon designed by the Montgolfier brothers carried two Frenchmen about three hundred feet over Paris. In 1804, the chemist Joseph Louis Gay-Lussac ascended four and a half miles without ill effect; but in 1875, two out of three men died of cold and suffocation when their balloon reached a height of six miles. In the 1930s, balloons with sealed cabins went far beyond this limit – one reached an altitude of thirteen miles; and in the 1960s, an un-manned balloon reached the height of twenty-nine miles.

In theory – to judge by its pressure of fifteen pounds per square inch – our atmosphere ought to be only five miles high. In actual-ity, it becomes thinner as it gets higher, so that at a hundred miles, air pressure is only a millionth of its value at sea level.

As our craft ascends into the troposphere – the lower portion of our atmosphere – the temperature drops steadily, and we soon

---

* Maxwell's demon began his interesting career by sitting at a tiny doorway between two containers of gas, opening the door to allow slow molecules to pass one way and fast ones the other. In this way, he defied the second law of thermodynamics – by allowing heat to flow from a colder to a hotter body – and illustrated Maxwell's moving-molecule theory of gases.

Crying for the moon? From William Blake's Gates of Paradise, *1793*

have to turn on the heating. At seven miles, we are suddenly buffeted by high winds that make our craft spin alarmingly. In fact, this westerly gale – known as the jet stream – blows at speeds up to five hundred miles an hour, and airline pilots take advantage of it. A mile further up, and all is calm again. Suddenly, we are floating peacefully in a region where the temperature remains constant: the stratosphere, the upper layer of the atmosphere. At this point we notice something odd. High up above us, there is a

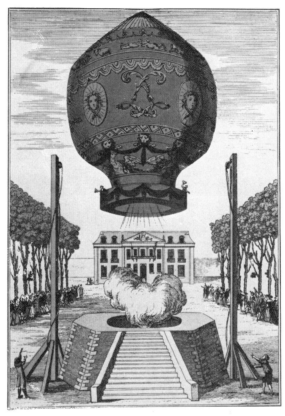

The ascent of De Rozier and D'Arlandes in the Montgolfier balloon, 21 November 1783

great silver cloud. But how can that be? We are miles above the earth's cloud layer. Then, suddenly, we are passing through it, and the interior of our craft is darkened. The demon explains that this is the remains of a vast dust cloud expelled by Mont Pelée in the great volcanic eruption of 1901; it was shot to a height of forty miles, and now, because of the fineness of its particles, it hangs suspended in the stratosphere, catching the light of the sun – the last trace of a tragedy that wiped out the city of St Pierre and destroyed over thirty thousand people.

Another ten miles and we are beyond the stratosphere – but still by no means beyond earth's atmosphere. The temperature is rising again, and the sun burns our faces. This is because we no longer have the protection of the atmosphere against ultra-violet waves, which cause sunburn. The demon remedies this by inserting a violet filter in the side of the capsule through which the sun is shining.

Our instruments register a height of seventy miles; our guide tells us that we are passing through the Heaviside Layer – one of the ionized layers of the atmosphere that reflects radio waves back to earth. It is, of course, quite invisible to us. But from this height, the view is incredible. The earth is blue and green with great white cloud masses, many of them in vortex patterns, and we can clearly see the Mediterranean – which is obviously enjoying a perfect summer's day – and North Africa. But now there is only blackness overhead, with the stars glaring like neon lights. At this height, they no longer twinkle – the twinkle seen on earth is caused by the atmosphere – and we

Apollo 15 takes off for the Hadley-Apennines area of the moon, July 1971

can also see that they have different colours. The moon is away on our left – looking, on the whole, much as it does from earth. The sun is far more blinding than on earth – even seen through the violet filter.

We are now at the height at which the first astronauts went into orbit around the earth – around a hundred miles. This, of course, was when they experienced weight-lessness for the first time. But we are still far from weightless: we seem to weigh about as much as on earth. The astronauts were weightless because they were travelling around the earth at 1800 miles an hour; in fact, they were really falling, but their speed kept them in a curve round the earth. Since we continue to travel gently upwards, magically defying the force of gravity, our weight remains much the same. It will only begin to diminish notably – by about one-tenth – when we are a thousand miles high.

Our guide decides that it is time to increase speed. The atmosphere is now thin enough not to worry about friction; yet its temperature has risen to over 500° – we are in the 'thermosphere'. There is a sensation like going up in a lift as our speed increases;

Above: The Aurora Borealis seen in Norway, 19 January 1839
Opposite: Zodiacal light seen in Japan in the 1870s
Below: The great meteor of 7 October 1868

suddenly, we feel very heavy. Three minutes later, our craft is plunged into darkness. Our guide explains that he has brought the radiation shields into operation because we are passing into the magnetosphere – or Van Allen belts – around fifteen hundred miles above the earth; here charged particles from the sun have been trapped by the earth's magnetic field – so that without the radiation shield, we would be subject to a bombardment of dangerous radioactivity. At this spot in the magnetosphere – which extends for thousands of miles into space – the radiation is intense enough to jam a Geiger counter.

Ten minutes later we are in empty space, beyond the last vestiges of atmosphere. However, no space is really empty, as we realize when there is a noise like rain against the walls of the capsule, and they are suddenly covered with tiny white stars, like impact marks on a car windscreen. We have passed through the edge of a meteor shower – the remains of a comet. Only the largest meteors reach the earth, burned up by our atmosphere; but space is full of these floating fragments and wisps of gas.

We are now losing weight by the minute – hardly surprising, since we are leaving the earth behind at a rate of ten miles a second. And soon we are floating gently and freely. It is a curious sensation – a languid feeling, as if there is no longer any hurry to do anything. You realize how much your tension and anxiety on earth is connected with your weight. (One of the original moon astronauts remarked about weightlessness: 'Contrary to being a problem, I think I have found the element in which I

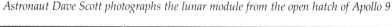

Astronaut Dave Scott photographs the lunar module from the open hatch of Apollo 9

belong.'¹) Our host produces sandwiches; you find it hard to take them seriously, because they are lighter than the lightest puff pastry. When you drink your coffee from a cup, you have to suck it out – there is no gravity to make it run into your mouth. Fortunately, the muscles of your digestive system exert such powers of contraction that there is no problem in swallowing.

Now that the earth is far below you, like a blue furry ball (that furry effect is caused by its atmosphere), you are at last beginning to get some idea of the distances in space. For the moon, although much brighter and clearer, hardly seems closer. And why should it? Even travelling at this speed, thirty-six thousand miles an hour, it is still six hours away; and if you were driving a car towards a mountain that was six hours away, you wouldn't expect it to get closer by the minute. The sun, blazing like a great white furnace in the blackness (it looks yellow on earth because our atmosphere filters out its blue light), is a three-month journey away. At this speed, even our nearest planetary neighbour, Venus, is a month's journey away. As for our most distant neighbour, Pluto, it would take us more than ten years to reach it, even at our present speed of almost a million miles a day. Our nearest neighbour in the starry universe, Proxima Centauri, would take a thousand lifetimes to reach.

Fortunately, Maxwell's demon is not subject to any of these limitations; he can take us anywhere in the known universe at any speed we care to name. His speciality is thought experiments. So I would suggest that, as we're already on our way towards the moon, we pay a call on our satellite.

As we draw near to that vast floating mirror, we note something that the first lunar astronauts remarked upon: it looks *three-dimensional*. We are so used to seeing it from the earth, looking like a dinner-plate in the sky, that it is surprising to see it turning into a real landscape, its craters rising above the plane like cathedrals. In 1822, the German astronomer Franz von Paula Gruithuisen startled the world when he announced that his telescope had shown him an immense city with giant ramparts. Better telescopes revealed that this was not a city, but simply a number of steep ridges. Still, looking at this twisted, pockmarked landscape, you can see how he made the mistake – and why William Herschel was convinced that the moon was inhabited. It looks like a war-ravaged landscape, full of giant shell craters. The Cambridge archeologist T. C. Lethbridge even advanced the interesting theory that there had once been a tremendous war between the moon and some other planet, and that these craters are the result of atomic bombardment. And indeed, they could be. That vast crater down there to the south – it is named after Newton – is five miles deep. That crater, below us, is Bailly, which is 183 miles across, and three miles deep. The whole satellite, with a diameter of just over two thousand miles, has over thirty thousand craters on its earth side (the moon always keeps the same face towards the earth).

What caused the craters? Some of them may have been volcanic; but the majority were formed through the bombardment of meteors. A meteor would crash into the moon at hundreds of miles a minute; a speed like that would send it to a depth of a mile or so. Then the heat and pressure would cause it to explode like a bomb – an atomic bomb – creating a hole into which you could easily drop a small country like Wales.

As we look out over this giant desert, we experience an odd sense of regret. It looks so impressive, more impressive than any scenery on earth; and it is so *useless*. No air, no water. How did it come to be here? A century ago, the most popular theory was that it was once a part of the earth, in the days when our planet was a molten mass of rock; its rotation caused a bulge, which finally split off, and floated away into space. . . .

Charles Darwin's son, Sir George, first put forward the idea, and an American astronomer named William Henry Pickering supported it because he thought that the roughly circular Pacific Ocean could have been the 'scar' caused by the moon's birth. This theory is now rejected, because the Apollo missions showed moonrocks to have a different composition from earthrocks. It seems more likely that when the planets 'condensed' out of some great dust-cloud, the moon formed just like the earth – close enough to it to be captured as its satellite.

A rather more exciting theory was put forward in the 1930s by the Viennese engineer Hans Hörbiger, who argued that our earth has had several moons, captured from the space between Mars and earth; the latest of them, the 'planet Luna', is also

The chemical composition of the moon's rock is slightly different from that of earth

the largest. Each of these moons has finally fallen to the earth, causing widespread catastrophes. Hörbiger was convinced that the moon was covered with a sheet of ice, 140 miles thick; he had been dead for nearly thirty years when the first Russian lunar module landed in September 1959 and proved that there is not the slightest trace of ice on the moon's surface. But Hörbiger *could* be correct about our moon's origin. If it was not originally part of the earth, then it could well have been 'captured'. Moreover, *if* previous moons had crashed on to the earth, they would have revolved faster and faster as they came closer, and all the seas would have been drawn into a great permanent tide round the equator – a tide that had no time to recede. As Professor Denis Saurat has pointed out, there *is* a strange line of marine deposits running from Lake Umaya, in the Peruvian Andes, and extending in a curve to Lake Coipasa, four hundred miles to the south, as if a 'tide' *had* once encircled the earth. . . .

The science fiction writer Isaac Asimov has an interesting theory about the moon; he feels it was a misfortune that the earth ever captured it.[2] Because ancient man could see the moon revolving around the earth, he naturally concluded that the rest of the universe did so too. This, says Asimov, led man to think of his earth as the most important thing in the universe, and of man, its ruler, as the most important creature. And here we are, he says, in the twentieth century, ravaging nature, destroying the environment – all because we believe that man is the measure of all things.

Think, says Asimov, what a difference it would have made if our moon had been captured by Venus instead. By observing its moon – which would have been quite visible from earth – the ancients would have realized that the morning and evening stars were one and the same. By observing that the moon appeared first on one side of Venus, then the other, they would have inferred that it revolved around Venus. It would have been logical to make the same observation about Venus and its position in

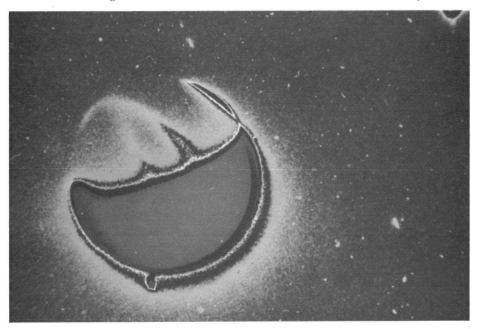

Ultra-violet view of the earth seen through the Apollo 16 telescope

relation to the sun, and to see that Venus (and Mercury) must revolve round the sun. Which means, says Asimov, that the Sumerians would have recognized that earth itself is probably a satellite of the sun, and the Copernican universe would have been recognized more than three thousand years sooner. And man would never have become so conceited about his position in the universe, and would not now be in danger of exterminating himself.

However, I cite Asimov's argument because it seems to me an interesting example of the 'rational' scientific outlook, with which this book takes issue. He assumes that man would have been less conceited – and destructive – if our ancestors had been aware that the earth was not the centre of the universe. Yet if Jaynes's argument is correct, man became conceited and destructive when he began to develop the bicameral mind; and this was because he had *lost* his intuitive sense of being the centre of the universe. Man is destructive because he is self-divided; and he is self-divided because he has separated his intuitive certainties from his rational thinking. In a paradoxical sense, man *is* a child of the gods and the centre of the universe; and if the

Copernican theory has taught him otherwise, then it has taught him something that is untrue. . . .

Still, surveying this bleak and meteor-scarred lunar wilderness, I have to admit that this point of view seems like wishful thinking. Man started to look at the stars because he thought the universe was full of interesting mysteries. Everything we learn about those remote worlds, however, deflates our expectations. The moon has a particular appeal to the human imagination; it has always been regarded as a source of mystery. In the early 1960s, I was much impressed by a book called *The Strange World of the Moon*, by V. A. Firsoff, in which he argues that there could be some form of life on the moon. He ends with the words, 'No, the moon is a mysterious and wonderful world, which may yet hide many surprises.' The Apollo landing of 1969 revealed that the moon is not at all mysterious, and that it does not really hold many surprises; it is simply dead.

But then, there may still be hope. When America's Orbiter-2 took photographs from twenty-nine miles above the Sea of Tranquillity in November 1966, they appeared to show the shadows of eight pointed spires, shaped rather like Cleopatra's Needle or the Washington monument. A Soviet space engineer named Alexander Abramov has stated that these objects constitute an 'Egyptian triangle', 'similar to the plan of the pyramids'.[3] And John O'Neil, a former science editor of the New York *Herald Tribune*, claims to have observed a giant bridge-like structure in the Sea of Crises, under which the sun shines at a low angle. No doubt these will prove to be natural formations; but until this is confirmed, we may as well cling to our romanticism.

A glance at the earth – which hovers in the lunar sky like a gigantic blue moon (four times as big as the moon as seen from earth) – reminds us that time is passing. Its rotation cannot be seen any more than we can see the movement of the hour hand of a clock, but the British Isles have already passed into its area of shadow. (Seen from the moon, the earth has its phases, just like the moon.) It is time we moved on.

The nearest planet to us at the moment – not counting the earth – is Mercury. Venus *should* be, of course, but at the moment, it is around the other side of the sun. So we may as well begin our survey of the planets in the logical manner with Mercury. There was a time when astronomers believed there might be yet another planet between Mercury and the sun; they even gave it a name – Vulcan. An amateur French astronomer, a Doctor Lescarbault, claimed to have seen it in March 1859, and was awarded the Légion d'Honneur on the strength of it. But the most rigorous search with modern cameras and telescopes has failed to discover it, and it seems fairly certain that Lescarbault made an honest mistake – perhaps seeing one of the asteroids, which had wandered out of its usual path between Mars and Jupiter. (This *can* happen, because the orbits are elliptical; Pluto, the outermost planet, actually ventures within the orbit of its neighbour Neptune.)

As we leave the moon behind us, we become aware once more how *battered* it looks. That little world had really been in trouble at some time in its history. And, in fact, we know roughly when it happened. About four eons ago – four billion years[*] – there was an incredible bombardment; the whole solar system became a shooting gallery, as one space scientist put it. The 'bullets' were chunks of rock shaped like billiard balls, each about thirty miles across. After that bombardment, the surface of the moon was covered with craters. Then molten lava boiled up from its depths, and created vast smooth areas – the so-called 'seas'. One of the stranger mysteries about this upwelling

---

[*] In this book, I refer to a thousand million years as a billion, rather than to the English million million.

*In astrology Mercury defines and enhances the quality of adjacent planets. It represents
the mutable part of man's nature. From* De Sphaera, *a fifteenth-century Italian manuscript*

is that it is more pronounced on the earth-side of the moon; so the earth was in some
way responsible. No one has yet worked out quite how.

We have crossed the orbit of Venus, approaching Mercury at half the speed of light.
But no one is very interested in Mercury; the most amazing object in the sky is the sun.
It is huge – more than twice the diameter seen from the earth – and expanding moment
by moment. But it is not so much the size that is impressive as the activity. From earth
the sun looks simply like a white-hot ball. Now we can actually *see* that it is a
monstrous furnace. Through the darkened walls of the capsule, we can see great
tongues of flame erupting and spurting far above its surface. The sun is actually *boiling*,

like a vast witch's cauldron; its surface seethes like the fat in a chip-fryer as the tremendous energies from below force their way out. Those spurts of flame – called prominences – actually leap up for hundreds of thousands of miles in great exploding geysers. They turn into fantastic shapes – trees, mushrooms, animals – so that it is like watching some extraordinary light-show. The great black areas called sunspots – which Galileo claimed to be the first to observe – look like vast cavities or craters; in fact, they are magnetic storms, each one thousands of miles square.

Because of the glare of the sun, we fail to notice Mercury until we are almost on it. The ancients had trouble seeing it for the same reason; it was so close to the sun.

Now, as we swoop towards it, it looks rather like the moon. In fact, Mercury is only slightly larger than the moon. Its surface is very similar, being pitted with huge craters. But as we draw closer, we see that it looks rather less battered than our moon. Its craters seem farther apart, and rather less big. The reason is that Mercury is a far heavier world than our moon; it has an immense iron core as big as the moon. So when it was subjected to the same terrific bombardment of 'billiard balls' four eons ago, the explosions made smaller craters. Moreover, as the iron core cooled and contracted, the surface became wrinkled, like a wizened apple. So lines of low cliffs run for hundreds of miles across its surface.

Until 1962, it was believed that Mercury had no rotation – that it always kept the same side turned towards the sun. This raised the possibility that there might be frozen gases on its dark side, remnants of some early atmosphere. Then, in 1962, radar revealed that the temperature of the 'dark side' was far too warm for it to remain permanently turned away from the sun. In fact, Mercury rotates on its axis very slowly, taking fifty-eight and a half days, and takes eighty-eight days to orbit the sun. The iron core means that it has a weak magnetic field, and it has an extremely thin atmosphere – almost a perfect vacuum. But even if there were creatures which could breathe it they could not live in that surface temperature of 700° Fahrenheit. If the moon is cold and dead, Mercury is hot and dead. So we decline our guide's offer to land in the Caloris Basin – a vast impact crater, over eight hundred miles in diameter, covered with smooth lava – and press on towards Venus.

The sun now fills the sky. If we were not in free fall, the sun's tremendous gravity would make us feel as if we had been turned into lead. It would be difficult to breathe. Gravity on the sun is twenty-eight times that of earth, so that I would end up weighing over two tons. And this explains why that vast furnace does not explode – its immense gravity holds it together.

It suddenly strikes us, with horror, that our guide intends us to dive straight into the sun. Before we can protest, we have plunged into the seething darkness of a sunspot. We cringe, waiting for the impact; but nothing happens. We also realize, to our surprise, that the sunspot is blindingly bright; it only appears dark in comparison with the rest of the sun. It is, in fact, part of the sun's magnetic field protruding into space, like a broken spring sticking through a mattress. Or you could think of it as being rather like a knot of wood in a tree. At all events, it prevents the energy from below from flowing smoothly to the surface.

It seems strange to be plunging inside a star, while being able to see as well as ever. The reason is that the sun is basically a gigantic atomic explosion – a hydrogen bomb

Opposite: (above) Solar eclipse showing prominences, seen by Warren de la Rue, 18 July 1860; (below) Solar prominences seen at Palermo by P. Tacchini, 8 July 1872

that, under tremendous pressure, converts hydrogen into helium; naturally, there is plenty of light.

Being inside the sun is a vertiginous experience, like being caught in several whirlwinds at the same time. On earth magnetism is found in solid bars of iron. Here on the sun it is associated with flowing gases. Consequently, the magnetic field is bent into all kinds of weird shapes; the magnetism at the sun's centre is like a *very* tangled ball of wool.

The craft has slowed down noticeably; evidently the sun's density *does* increase towards the centre. Yet at no point does it approach being solid – the temperatures here are too great. And this, incidentally, is the part of the sun that Herschel thought might be cold, and harbour living creatures. . . .

So no crash comes; and, travelling at half the speed of light, we emerge from the other side of the sun – its diameter is about 860 000 miles – in about ten seconds.

Now we are in space again, comfortably shielded from the blinding glare behind us by the opaque filter. As our eyes accustom themselves to the velvet darkness of space, we see Venus, the most brilliant and impressive object in the heavens. Even from earth, Venus is the most noticeable body in the evening and morning skies, the inspiration of many love poems. Now it hangs in front of us, about the size of the moon, dazzling. It seems to shine with a pearly inner-light. This, of course, is an illusion; it shines by reflected sunlight, like the moon. The difference is that this is ten times brighter than our moon, which only reflects 7 per cent of the sunlight; Venus reflects 76 per cent. (So we say that it has an albedo of 76 per cent.) The brightness is caused by the fact that Venus is entirely covered with heavy clouds, and these reflect the light. For some reason, this light is distinctly yellow.

Up to a few years ago astronomers liked to think of Venus as the 'earth's twin'. It is, in fact, almost the same size. It has an atmosphere with an abundance of water vapour and carbon dioxide, but no free oxygen. So the picture of Venus that emerged around the beginning of the twentieth century was of a planet that might look much like our earth, under its cloud blanket. Since the clouds (which are below freezing point, and therefore made up of ice crystals) deflect more than half the sun's light and heat, temperatures on Venus might be approximately the same as on earth. In 1908, the Swedish scientist Svante August Arrhenius wrote a book called *Worlds in the Making* in which he argued that life existed wherever it could find a foothold in the universe, and that Venus was probably much like our earth in the Carboniferous Period – complete with oily swamps, huge conifers and early dinosaurs. Previous speculators, including Swedenborg, Kant and George Adamski – who claimed to have been there in a flying saucer – peopled Venus with man-like creatures. But radio measurements of Venus made in 1956 revealed that its temperature was too high for liquid water to exist; it seemed to be at least 600°F. That also punctured a theory, put forward in 1954, that Venus was entirely covered by a sea.

On 27 August 1962, a spacecraft called Mariner II left earth for Venus; it swept past the planet in December. The messages sent back exploded all the romantic theories about Venus's surface. For it proved to be even hotter than the radio emissions had shown – about 900°F, hot enough to melt lead. Then, between 1965 and 1978, the Russians succeeded in landing no less than ten space probes on Venus, and two of these sent back photographs. These, predictably, showed barren landscapes.

As we approach the upper atmosphere of Venus, the windows of the capsule frost up. But as we float down, the frost vanishes and we move into a yellowish mist. This is sulphuric acid gas, which absorbs moisture. Finally, we emerge from the yellow clouds into a thin fog. We peer down through this, straining for a glimpse of the surface; but we can see nothing. This, our guide explains, is not because the atmosphere is full of mist, but because it is so dense that light is scattered or diffused. Blue light is scattered more than red because its waves are shorter. That is why distant mountains look blue and hazy; it also explains the blueness of the sky. Here the whole atmosphere is penetrated with that blue mountain haze. Accordingly, our guide hands us red goggles. As soon as we put these on, we can see better. But there is still not much to see. The surface below us looks flat and indistinct. It is periodically illuminated by tremendous flashes of lightning; in fact, we seem to be in the middle of an enormous thunderstorm. The clouds of Venus are full of static electricity, and the lightning never stops. Our craft is battered by a wind which seems to be the Venusian equivalent of earth's jet-stream. A temperature gauge on the wall shows that it is *very* hot out there; while even at this height – thirty miles or so – air pressure is as great as on the surface of the earth. As we continue to descend, this rises steadily.

And now we can begin to make out a few features. There are mountains, but not half as spectacular as those on the moon. There are also some craters, but these look as if they have been formed by volcanic action; the atmosphere would protect Venus from most meteorites.

At last, we touch down on the surface. When we push up our red goggles, we can only see thirty yards ahead; it is like a misty morning on earth. There is a heavy twilight around us, and things seem to shimmer and loom as if seen through water. In a sense, we *are* seeing things through water, for the air pressure here is as great as if we were more than half a mile under the sea; as a result, the atmosphere – mostly carbon dioxide – is much 'thicker' than on earth. Even without the goggles, everything looks distinctly red, because the upper atmosphere has filtered out most of the blue light. This rosy light is not unpleasant – it gives the place a romantic air, rather like a sunset on earth. When we re-adjust the goggles over our eyes, we can see we are on a rocky plateau, with mountains in the distance. It looks somehow less bleak and unfriendly than the moon or Mercury. As our capsule floats forward again, a few feet above the surface, we realize that this is because the landscape is less scarred and weathered. There are no signs of the howling gales that astronomers used to predict, no great dust clouds; everything seems to be still, except for the flashes of lightning.

Again, no sign of life – nothing at all. Occasionally, we think we see signs of algae on the rocks, but it turns out to be an optical illusion. This place is full of such illusions. After half an hour of floating around in this red twilight, it becomes oppressive. Finally, we take off again, moving upwards in a long, oblique line. And before we are half a mile high, we see something that dispels the idea that everything on this planet is on a small scale. An immense rift valley opens out below us. It has sheer cliffs which plunge down to a depth of three miles or so, and it stretches as far as the eye can see in either direction. Yet this Venusian equivalent of the Grand Canyon – 900 miles long and 175 miles wide – has not been formed by a river. The surface has simply cracked apart, like the lips of a man dying of thirst. This hot, dry planet, with its sulphureous atmosphere, has more than a touch of hell. In fact, the planetary scientist Carl Sagan points out that ancient legend seems to associate the place with hell, since Lucifer is called 'the son of the morning' – the morning star. It is an interesting thought – that our remote ancestors may have derived their idea of hell from some dream-like intuition of conditions on Venus!

As it reflects the sun, the moon is held to determine
man's **persona** *or conscious mind, the personality which masks*
the true self. From De Sphaera

·SOL·

*The sun, on the other hand, represents man's essential
being. Its position in the zodiac determines an individual's
basic character and fate. From De Sphaera*

It is a relief to pass through those sulphureous upper vapours and out again into space, whose blackness and cold seem almost welcoming.

Now we are moving on towards Mars, the planet that has intrigued and fascinated human beings more than any other. Why should this be so? The ancients associated Mars with war, ambition and ruthless energy. The reason may be simply that Mars looks red. But the statistical researches of the Gauquelins into astrology showed a convincing correlation between people born 'under' Mars, and various kinds of sporting activity. In *Cosmic Influences on Human Behaviour* Michel Gauquelin gives a list of key words associated with various 'planetary temperaments'; those associated with Mars are words such as 'active, eager, quarrelsome, reckless, combative, courageous, dynamic, self-willed' and so on. (It must be emphasized that Gauquelin was not trying to 'prove astrology', and that he still maintains that most astrological beliefs are nonsense; his results with sportsmen have been duplicated by a Belgian committee of scientists led by an astronomer.)[4] These words obviously apply equally to warriors; and in most ancient cultures – Greece, for example – these two groups would be one and the same. So in the case of Mars, it could be argued that there is certain basis for the traditional ideas about its influence.

Galileo turned his telescope on Mars in 1610, looking for the 'gibbous phase' (that is, the phase when it is 'three-quarters-full'), but his telescope was too weak to reveal it. Francesco Fontana saw it in 1638, and his sketches of Mars may be described as the first inkling of the modern Mars mystery. They showed a large dark spot in the middle of the planet. When Christian Huygens turned his more powerful telescope on it in 1659, he saw no such dark spot; but he *did* see the ice cap at the south pole. If a planet had an icy south pole like earth, it seemed reasonable to assume that it would have seas like earth, and would therefore also have a weather system much like our own. Huygens remarked that some parts of Mars were darker than others, and refers several times to 'its inhabitants'.

Cassini's nephew Giacomo Filippo Maraldi observed the planet for thirty years, and concluded in 1702 that the markings on Mars showed signs of change over the years; this was particularly noticeable at oppositions (the time of the planet's closest approach to the earth, roughly every 780 days). So it seemed possible that Mars had large areas of vegetation that changed with the changing seasons. In the 1780s, William Herschel published two papers on Mars in which he commented on its changing colours, and decided that 'its inhabitants probably enjoy conditions analogous to ours'.[5] Yet no one was excited by this thought; most people took it for granted that the moon and planets were probably inhabited, just like our earth.

During an opposition in 1858, Pietro Angelo Secchi in Rome made the first colour drawings of Mars; five years later, Lockyer did the same. And in 1867, Richard Anthony Proctor drew a map of Mars with various features marked as continents and seas. Since he chose mostly English names for the features, his map caused roars of rage among European astronomers; the hostility generated may account for the neglect of his pyramid theory three decades later.

Then, in 1878, came the map that suddenly made Mars a universal talking point – rather like the Loch Ness monster in the 1930s. It was drawn by an Italian, Giovanni Virginio Schiaparelli. And what caused the excitement were the thin dark lines that Schiaparelli drew between the 'seas', and which he called *'canali'*. In Italian, this simply means channels or grooves; but popular journals preferred to translate it as 'canals'. In fact, the dark lines on Schiaparelli's map were so thick that the 'channels'

Martian invaders killing humans with their death rays.
From H. G. Wells' The War of the Worlds

they depicted would be a hundred miles wide. It made no difference. As far as the man in the street was concerned, Mars had immensely long canals which had been constructed by intelligent beings. And since terrestrial engineers had had enough trouble building the Suez canal, a mere 107 miles long, it followed that beings capable of building thousand-mile waterways must be far ahead in scientific achievement.

It was *this* that captured the imagination. If someone had suggested that Mars was inhabited by Australian aborigines, no one would have been interested; 'civilized' men thought so little of aborigines that they had exterminated the inhabitants of Tasmania. But technologically developed beings were a different matter. Man has an innate longing for something he can look up to and admire – for beings who can solve his problems and show him the way to self-improvement. Suddenly, there were

Left: Sunrise on Venus, taken by the Pioneer orbiter spacecraft on 5 December 1979
Right: Bright polar rings and cloud markings on Venus, taken from 40 000 miles

dozens of books about Mars and its canals and cities and strange inhabitants. 'Those who have never seen a living Martian can scarcely imagine the strange horror of its appearance,' wrote H. G. Wells in *The War of the Worlds* in 1898, 'The peculiar V-shaped mouth with its pointed upper lip, the absence of brow ridges, the absence of a chin. . . .' Professor Jakob Schmick of Cologne contributed *Mars, A Second Earth* (1879), while an Austrian, Otto Dross, explained in *Mars, A World Engaged in the Struggle for Survival* (1901) how the canals were necessary to conserve the small amount of water left on the planet. A French widow named Clara Goguet offered a prize of 100 000 francs to the first scientist to find a way to communicate with any world *other than Mars*, apparently convinced that this could happen at any moment. Ludwig Kann argued that Mars was covered with sea, and that the dark patches were vast islands of floating seaweed. Hans Hörbiger, whom we have already mentioned, argued that Mars was covered with thick ice, and that the canals were cracks in the ice. Svante Arrhenius argued that Mars would be far too cold to sustain life.

In 1895 an American astronomer published his first book on Mars; he was an aristocratic Bostonian named Percival Lowell. Lowell had been thirteen in 1878 when Schiaparelli brought out his study of the canals of Mars; he was immensely excited, and decided that one day he would make his own contribution to the solution of the mystery. Since he was wealthy, Lowell was able in time to realize his dream by building an observatory in Flagstaff, Arizona, where the air was clear. Staring at Mars through his powerful 24-inch refracting telescope, he was convinced that he saw the canals, and that they were too straight to be natural markings. Mars had to be

Mars from around 200 000 miles. The huge Valles Marineris is towards the top of the frame. The south pole is in the dark, bottom left

inhabited. And that, after all, was not such a far-fetched hypothesis. Mars is further away from the sun than earth – by some forty million miles. So it must have cooled down sooner. But assuming that it was formed out of the same cosmic material as earth, it must have passed through the same stages as our own planet: hot seas, gradually cooling; then life forming in those seas. And finally, the evolution of intelligent beings, who became engaged in a struggle for survival as their planet gradually lost its atmosphere and water – for Mars has only one third of earth's gravity – and were forced to create canals.

It was an exciting scenario, and Lowell's respectability as a scientist made it more convincing. (In appearance he was so naturally aristocratic that photographs of him in foreign editions of his books are sometimes labelled 'Sir Percival Lowell'.) If other scientists failed to see his canals, it could have been because their telescopes were not as powerful as his, or that their observatories were less favourably situated.

Yet we should try to grasp the difficulties involved in seeing Mars from earth. It is a tiny planet, with a diameter only twice that of our moon. But our moon is a quarter of a million miles away; Mars, at its closest, is thirty-six million miles. So we have to imagine our moon at more than seventy times its present distance. Obviously, it would be very small – so small that our most powerful telescopes would barely reveal its craters. (The existence of craters on Mars was suspected from the 1890s onward, but not confirmed until the first Mariner flight in 1965.) Lowell was, in fact, totally mistaken about the canals. He was seeing small individual markings on the surface of Mars, like our moon's craters, and his eyes were 'joining them up'.

Yet even when astronomers came to reject the canals as optical illusions – once telescopes more powerful than Lowell's failed to show them – the notion of life on Mars persisted. Most of the books on astronomy in my own library published before 1965 seem to agree that there is probably some sort of vegetation on Mars, if only in the form of lichen. And the intelligent-life hypothesis continued to have its adherents; the Soviet space scientist I. S. Shklovsky has suggested that the two moons of Mars – discovered by Asaph Hall in 1877 – could be artificial (or Martian-made) satellites.

All these dreams collapsed in July 1965, when Mariner 4 swept past Mars at a distance of approximately seven thousand miles, and sent back to earth twenty-one television pictures of its surface. They showed a surface very like the moon's – barren and pitted with craters. Subsequent Mariner probes revealed 'chaotic' areas of jumbled ridges and valleys, and smooth areas that could be deserts, or else lava-covered 'seas' as on the moon. Finally, in July 1976, the American probe Viking 1 landed on Mars and transmitted its pictures back to earth. They showed a bleak, bare landscape, reddish in colour, and covered with small rocks. The soil was tested – with a kind of chemical 'soup' – for any sign of life. All experiments but one were negative. The exception – whose result is ambiguous – showed some kind of reaction taking place; but this could have been purely chemical.

What seems clear, then, is that all those romantic expectations about life in the solar system are doomed to disappointment. For if there is no life on Venus or Mars, then there is pretty certainly none on Jupiter or Saturn, or any of the more distant planets.

Still, we have to admit that, at close quarters, Mars is the most interesting planet we have seen so far. It is a globe of many colours – predominantly red and yellow, but with touches of blue, green and white. The white is 'snow' – frozen carbon dioxide. And as we float down over the northern pole, the scenery looks very much like earth – it could be the Tibetan plateau. At this distance, we can see that Mars is not really like the moon or Mercury. There are, admittedly, many craters; but they are far more eroded. Earth itself once had thousands of craters; but our weather system is so active that they have mostly been obliterated by wind, rain and glaciers. Mars also has a vigorous weather system, as we realize when driving 'snow' covers one surface of the capsule. Elsewhere on the planet, these same winds raise immense dust clouds, which cover most of the surface for days. And the storms act on the scenery like a great sand-blaster, cutting into the rock. The result can be seen as we leave the winter region of Mars behind us, and cross a desert area.

The rock formations below are eroded and fantastic, as in desert areas of the earth; there are immense cliffs showing layers of geological strata, and solitary pinnacles of rock rising out of the sand. As we pass close by, we can see that the scale of everything is greater than on earth. It seems absurd, on a planet so much smaller than ours; yet there are mountains whose bases are three hundred miles wide, ridges that rise more than ten miles into the air, canyons that make the Grand Canyon look like a modest gully. If we wanted to be moralistic, we could say that Mars is what the earth might become if we continue to destroy the environment. The Sahara desert was a green, fertile region five thousand years ago, in spite of changing wind systems which (after the last ice age, eleven thousand years ago) brought less and less rain. Then man

Opposite: (above) Percival Lowell in his library; (below) Lowell's drawing of Mars, 1903
Overleaf: Voyager 1's view of Jupiter. Io can be seen against the planet's disc. On
the far right, showing as a white spot, is Europa

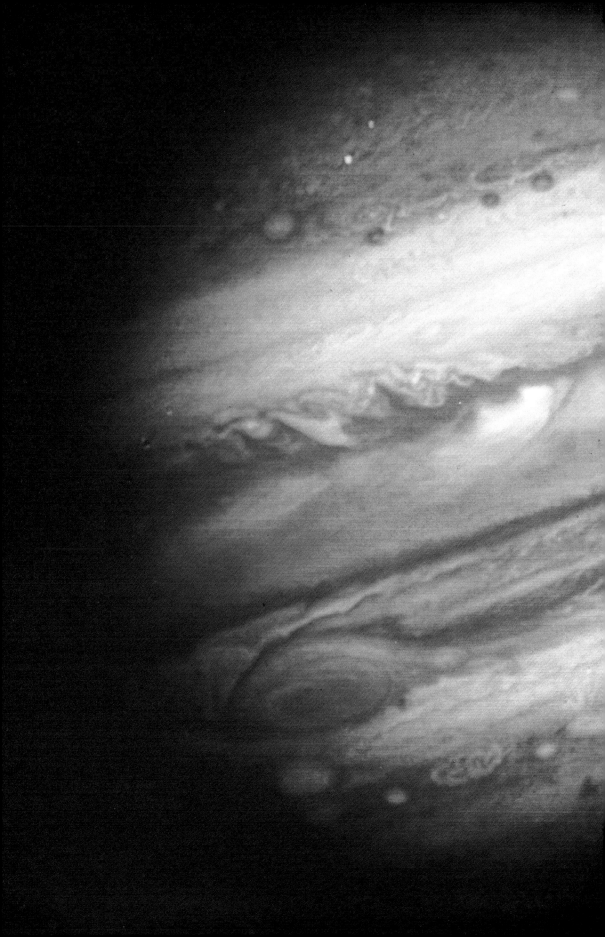

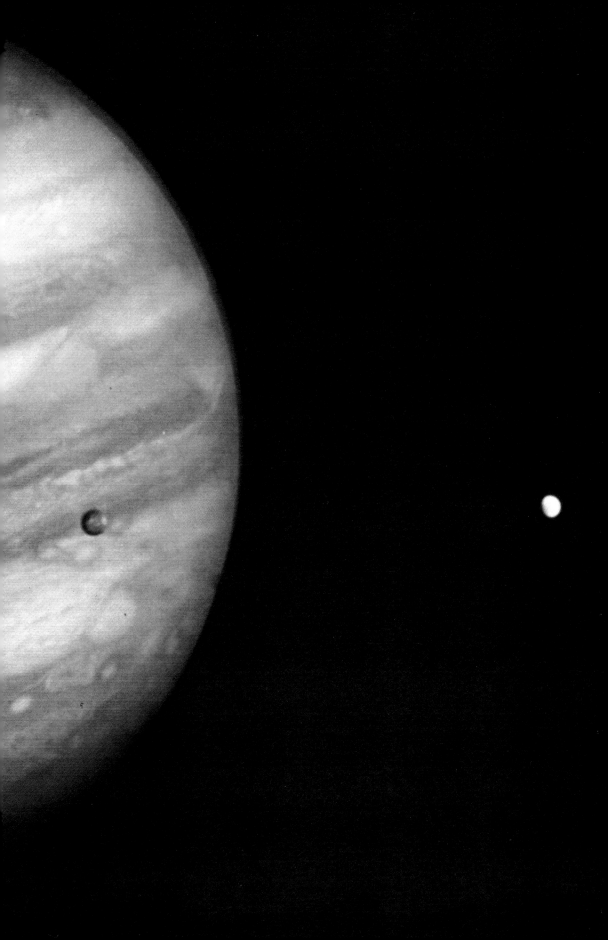

·MARS·

*Mars, the fiery red planet, embodied both positive and negative military virtues
in Babylonian, Egyptian and Greek astrology. From* De Sphaera

allowed his animals to overgraze it, and hacked away forests to make roads; until the
balance finally turned against life. On Mars, a similar process occurred without – as far
as we know – the help of living creatures. It shows clear evidence of old watercourses.
But it appears that the temperature dropped; the water froze in the soil; and the decline
became irreversible. Without water to sculpt its landscapes with rain, plane away its
volcanoes with glaciers, and fill its valleys with mud, Mars turned into a giant
dust-bowl, with howling winds and flying sand eating into its crust.

How recently did all this happen? We cannot be sure. Carl Sagan speculates that
Mars could have possessed rivers in the past thousand years or so; it could have been

five million. *If* Mars was once very much like earth, then it confronts us with the question of why it seems to be lifeless. Fred Hoyle has argued that life may have arrived on earth on meteorites; if so, one would expect it to arrive on Mars in the same way. So there ought, somewhere on that barren planet, to be fossils of long-dead life forms – perhaps even the remains of great cities. There *could* still be life, locked away under its sands, waiting for the return of more clement conditions – Mars is at present in the midst of an ice age – to reappear. . . .

By now we have travelled a thousand miles over the surface of Mars, and it is clear that there is little more to see. The scenery can be spectacular; but then there are hundreds of miles of the boulder-strewn plains which are known in the Sahara as Hamadas, and the gravelly deserts called Regs. At some future date, man may be technically advanced enough to recover the carbon dioxide from the rocks of Mars, and bring plant life to small areas. Until that happens, Mars will remain a graveyard.

As we leave Mars, we pass within a few miles of its largest satellite, Phobos. And it is clear that Shklovsky was mistaken. This is not an artificial satellite: just an irregular chunk of rock, about twenty miles long. It is not even roughly spherical. One of these days, it will crash onto the surface of Mars, like its smaller companion, Deimos, and become just another part of that endless rocky desert.

We have now visited the three inner planets of the solar system, and there are still five to go. But these five lie at increasingly vast distances. In order to get some idea of the scale of the solar system, let us borrow a convenient illustration from Sir James Jeans. Imagine that the solar system is the size of Piccadilly Circus (or Times Square). In the centre is the sun, the size of a pea. A mere nine inches from the pea is Mercury – a pinhead so tiny that you would need a magnifying glass to see it. At fifteen inches from the sun, there is Venus, just visible to the naked eye. At nearly two feet from the sun there is the earth, another just-visible pinhead. Ten inches further on, we encounter Mars. We are still less than three feet away from the centre of Piccadilly. And to get to the next planet, massive Jupiter – about the size of a grape seed – we have to move ten feet from the sun. To reach Saturn, which is only slightly smaller than Jupiter, we have almost to double that distance. To reach Uranus – half the diameter of Saturn – we have to travel another twenty feet. Out to Neptune, another twenty feet. Finally, the voyage to tiny Pluto – smaller than Mars – would be about another fifteen feet. (All these are mean distances – we have to remember that planetary orbits are egg-shaped.) So Pluto is about seventy-five feet from the sun, as compared with the earth's two feet. Even a beam of light takes around twelve hours to travel from the sun to Pluto – as compared with about nine minutes to the earth. Clearly, then, all these 'superior planets' are too far from the sun to have life as we know it. If Mars is too cold at three feet from the sun, what chance is there for Jupiter, at ten feet?

But before we reach Jupiter – where we have been preceded by four space probes – we have to run the gauntlet of that rock-strewn Sargasso Sea of space called the asteroid belt. The asteroids are chunks of rock, not unlike those moons of Mars (which may originally have been asteroids); they are probably the building-blocks of a tiny planet which never assembled itself, midway between Mars and Jupiter. Kepler felt that the gap between the two planets – a leap from three feet to ten in our Piccadilly analogy – was a little too great, and that there ought to be another planet there. And Bode's Law predicts a planet at about 250 million miles from the sun. Yet the space seems empty. At least, it did until New Year's Day 1801, when Father Giuseppe Piazzi

Callisto, photographed by Voyager 2 at a range of 677000 miles

of Palermo discovered a 'comet' in the constellation Taurus. Oddly enough, a group of astronomers, led by Baron Franz Xaver von Zach and Heinrich Olbers, had just decided to form a 'zodiac committee' to search for the 'lost planet' between Mars and Jupiter, and Piazzi had been invited to join. It soon became clear that Piazzi's 'comet' *was* the planet – or one of them. For in March 1802 Olbers relocated Piazzi's planet – now called Ceres – and then discovered another one nearby. This was named Pallas, and Olbers immediately suspected that he was looking at fragments of a planet that had disintegrated. Professor Huth of the University of Dorpat had another suggestion: that these fragments were the pieces that might one day form a planet, as once the earth had been formed from floating fragments. Huth predicted that more would be found, and Olbers soon found two more. Ceres, the largest, was six hundred miles in diameter. As new asteroids were discovered every other year, it became clear that there were hundreds of them; in fact, in 1891, Max Wolf discovered no less than 228 asteroids by means of the camera. At the latest count there are about thirty thousand.

Why are there so many fragments? One theory is that the gravitational field of Jupiter prevented the material of a planet from coalescing; so it has remained scattered in space. But if it *was* once a planet, what knocked it to pieces? Could it once have been inhabited, and been destroyed in an atomic explosion? The present state of knowledge suggests that we may be in the dark for a long time.

Out beyond the asteroids lies Jupiter, the greatest planet in the solar system, with a diameter one tenth of that of the sun. When I was a schoolboy, my weekly comic had a serial called 'The Jolly Giants of Jupiter'; and astrologically speaking, Jupiter *is* a jolly giant – a planet associated with good luck. Unlike the asteroids, however, it has no exciting history of discovery; Jupiter is so big that it has always been known. But when the telescope was invented, new discoveries were made almost daily. Powerful telescopes revealed cloud belts parallel to its equator – belts that changed day by day. This is explained by the fact that Jupiter positively whirls on its axis – its day is just under ten hours; and since its surface is so vast, this means that anyone living on its equator would be carried along at twenty-seven thousand miles an hour (as compared to a thousand miles an hour on earth). Yet its gravity is less than three times that of earth. This is because Jupiter has an extremely low density – a mere 1.34 times that of water. Richard Proctor guessed correctly that it was a great bubbling mass of fluid. In 1878, the German astronomer Ernst Wilhelm Lieberecht Tempel observed the immense Red Spot, a great ruby-coloured oval just below the equator; its length was estimated at thirty thousand miles, so the earth could easily be dropped into it. (Cassini may have noticed it two centuries earlier.)

The only thing that seemed certain about Jupiter in the nineteenth century was that it was a highly disturbed planet – a kind of miniature version of the sun. (Though far colder – it has been described as a porridge of freezing gas.) The two Pioneer space probes of 1973 and 1974 confirmed this view. The Red Spot is a vast anticyclone; its red colour comes from red phosphorus; and it is slowly shrinking – it may well not exist a century from now. There are other great areas of disturbance which come and go – one vast storm lasted throughout most of the first half of this century.

So as far as our present voyage is concerned, there would be no point in trying to land on Jupiter. There would be nothing to land on; we would simply pass through increasingly dense layers of gas. A few thousand miles down, we would swim through liquid hydrogen. Eventually, about forty thousand miles down, we would encounter a molten core, probably of iron silicate. The whole seething mass radiates heat, and is surrounded by Van Allen belts far more powerful than the earth's. If we felt inclined to risk venturing beyond its roaring envelope of gases, we would need a speed of thirty-seven miles per second to escape again.

The satellites of Jupiter show an interesting variety. The largest and outermost, Callisto, is over three thousand miles across and is the most heavily cratered body in the solar system. Europa is as smooth as a billiard ball – possibly because its icy surface flows to cover the scars caused by meteorites. Ganymede is made of ice overlain with rock, cracked and cratered. And Io is an enigma: a solid chunk of rock, seething with volcanic activity. To add to the complications, Voyager 1 revealed that Jupiter also has a 'ring', like Saturn, too faint to be seen from earth.

Let us press on to the next giant of the solar system, Saturn. This involves going almost twice as far as we have already come; not entirely a pleasant experience. There is something worrying about seeing our sun looking like a small ball-bearing, shining like an unusually bright star, instead of blazing like a furnace. It makes us realize how

Venus, if well placed in a horoscope, influences our ability to love, attract, and express
ourselves artistically. If badly aspected, she may distort the emotional nature

much we take the sun for granted on earth. Leaving it behind like this produces a strange, deep feeling of insecurity. Pascal was right; there *is* too much space. . . .

As for Saturn, it is difficult to work up much enthusiasm for that grey wanderer. The astrologer William Lilly, writing in 1647, says, 'He is not very bright or glorious, nor does he twinkle or sparkle, but is of a wan or pale ashy colour. . . .'[6] The ancients, quite frankly, regarded Saturn as a planet of bad luck. Modern astrology takes a more optimistic view, regarding Saturn as the bringer of responsibility and maturity. (In Holst's suite *The Planets* Saturn is the bringer of old age.) Gauquelin's statistical researches showed that scientists tended to be born 'under Saturn', and that typical

·SATVRNVS·

The outermost planet known to the ancients, Saturn symbolizes dispassionate intellect.
The Greeks identified it with their old deposed ruler of the gods, Kronos

traits associated with 'Saturnians' were caution, restraint, taciturnity and reliability.

Because of its immense distance – around eight hundred million miles from earth – and the relatively small amount of light it receives from the sun, we know very little about Saturn. Its density is so low – about .7 that of water – that it seems a reasonable assumption that it is another ball of gas like Jupiter; its main constitutent seems to be hydrogen. But because of its size – its equatorial diameter is seventy-five thousand miles, more than nine times that of our earth – the hydrogen a few thousand miles below the surface would be as solid as metal; the core itself probably *is* made of metal, much like that of earth.

*Saturn and its rings as seen by L. Trouvelot, 30 December 1874. The 'rings' are
possibly the remains of a satellite broken up by tidal forces*

At close quarters, Saturn proves to be an impressive sight – this is verified by the photographs sent back by the space probe Pioneer 11, which flew close to the planet in September 1979. One report speaks of a translucent, marble-like surface graced with diaphanous rings'. The famous rings, discovered soon after the invention of the telescope, are really millions of tiny satellites – chunks of rock and ice floating around Saturn, like a miniature asteroid belt. They extend more than fifty thousand miles out into space, and absorb radiation. (So Saturn has no Van Allen belts.)

Saturn has eleven moons – one of them discovered by Pioneer 11 – the largest of which, Titan, is larger than Mercury. Space scientists once entertained hopes that Titan might support some form of life, since it is large enough to possess an atmosphere. But Pioneer 11 dashed that hope when it recorded that the upper atmosphere has a temperature of 392° F below zero – too cold for any known form of life. So the marble-grey planet which is as light as a beach ball is cold and dead. Admittedly, it radiates a certain amount of energy; but this is probably caused by slight contraction inside.

On, then, to Uranus, Holst's 'magician', the planet whose discovery made William Herschel famous. This planet is only half the size of Jupiter, although still nearly four times the diameter of our earth. Its chief peculiarity is that its axis of rotation is almost in the plane of its orbit – that is to say, Uranus lies 'sideways'. At the time I write this, Voyager 2 is on its way out towards Uranus (due to arrive in 1986). But no one expects spectacular discoveries. We know Uranus has rings, like Saturn – they were discovered in 1977. It also has five satellites, although at such immense distances – Uranus is never less than 1600 million miles from earth – it is difficult to judge even their size. Spectroscopic analysis suggests that Uranus is basically a smaller Saturn, with hydrogen and methane. Yet the planets of our system have turned out to be so interestingly different from one another that I, for one, shall be surprised if Uranus turns out to be a mere duplicate of any other planet.

Certainly, Uranus differs slightly from its neighbour Neptune, in spite of the fact that they are roughly the same size. To begin with, Neptune seems to have some kind of internal heat source. Neptune also has a 'normal' axis, inclined at about the same angle as that of the earth. It has only two satellites – although, as we shall see in a moment, it may once have possessed a third. But it certainly shares one characteristic with the greater outer planets: it is basically a ball of gas, probably with a solid core, and is too cold to sustain life.

From our point of view, the most interesting thing about Neptune is the story of its discovery. This was a direct consequence of the discovery of Uranus by Herschel in 1781; but it was not discovered by the telescope. Two mathematicians tracked it down, like detectives hunting for a body.

In 1821, forty years after the discovery of Uranus, a French astronomer named Alexis Bouvard published tables describing its orbit, and noticed that it failed to obey Newton's inverse square law. It wandered away from its plotted path. It was possible that Newton's law could no longer be rigorously applied to a body so remote from the sun. But it was also just possible that there was yet another planet, so far undiscovered, lying beyond the orbit of Uranus and affecting its movements.

Twenty years later, in 1841, a young Cornish mathematician named John Couch Adams decided that he would apply himself to the problem of Uranus's disturbed orbit as soon as he had taken his degree at Cambridge. Two years later, he obtained a fellowship, and settled down to find an unknown planet with a pencil and paper.

The first thing he needed was the latest data on the orbit of Uranus, so that he could pinpoint the discrepancies – the points where its path deviated from the ideal orbit predicted by Newton's law. These were willingly supplied by Sir George Biddell Airy, the Astronomer Royal. Adams plunged into his calculations, and completed them in a matter of months, by September 1845. He had assumed that the unknown planet would be in the orbit predicted by Bode's Law; he was mistaken – Bode's Law breaks down for the two outermost planets. Yet his calculations based on Airy's figures were so precise that he was able to predict where the unknown planet could be located on 1 October, one month later.

And now the young mathematician encountered an incredible run of bad luck. His professor at Cambridge, James Challis, who had helped him obtain the information from Airy, failed to grasp the importance of the prediction. If Adams had written a letter to *The Times*, a hundred astronomers would have trained their telescopes on the spot he predicted. Instead Challis advised him to send his calculations to Airy. Airy was abroad. Adams called several times, without success – on one visit, Airy was having supper, and his servant refused to disturb him. Adams left his paper. Airy read it, and instead of pointing his telescope at the sky – unfortunately, 1 October had already passed – he began checking it for errors, and made various comments on its assumptions.

By one of those curious coincidences that happen so often in the history of science, a young Frenchman named Urbain Jean Joseph Leverrier became interested in the problem at the same time; he was encouraged by the director of the Paris Observatory, Dominique Arago. Leverrier started by studying Bouvard's ephemeris, and then examined the later figures; they showed that the 'discrepancies' increased as time went past – a fairly firm indication that another planet was to blame. In 1845 – the year

Adams had 'pinpointed' the unknown planet – Leverrier published a paper in *Comptes Rendus*, discussing these irregularities, although at this stage he did not speak about the possibility of an unknown planet. But in June 1846, he published another paper in which he stated that nothing could explain the irregularities of Uranus except the pull of a planet beyond it. He specified where it would probably be found – his position agreed within one degree with Adams's calculations. Airy began to correspond with Leverrier, but still failed to mention Adams.

The search was now on; but how would an astronomer distinguish the new planet from a star? He would need a precise star chart of that section of the sky. And neither the French nor the English possessed such a chart. One *did* exist – it had just been made by the thorough and systematic Germans – but they had not yet mailed it off to other observatories. The result was that it was the German astronomer Johann Gottfried Galle, at the Prussian observatory in Berlin, who was the first to turn his telescope on Neptune.

In England, Airy had asked Adams's professor, Challis, to look for the new planet with the Cambridge telescope. Challis took his time; he actually saw Neptune on several occasions, but failed to realize what he was looking at. Later he confessed that he did not believe Adams *could* be so accurate.

Sir John Herschel now stepped in and tried to gain Adams the recognition he deserved. He wrote a letter to *The Athenaeum* describing what had happened. The French were understandably indignant, and one French newspaper announced that it had uncovered a plot by Herschel, Challis and Airy to try to steal the credit for the discovery.

In due course, the scandal died down, and Adams received the credit due to him – books on astronomy invariably state that Neptune was discovered by Adams and Leverrier. Adams became a professor of astronomy at Cambridge, and at the age of forty-one succeeded Challis as the director of the observatory there. He was even offered the post of Astronomer Royal when Airy retired, but turned it down; he also rejected a knighthood. So the story ended more or less happily, and Adams led a long and fruitful life as a brilliant mathematical astronomer. Challis, the cause of all the trouble, seems on the whole to have escaped severe censure.

Not long after the discovery of Neptune, the English astronomer William Lassell discovered that Neptune had a satellite; it was named Triton. This seems to be as big as Saturn's moon Titan – which is larger than Mercury. But the rings Lassell thought he saw round Neptune have not been verified by subsequent astronomers. A second small moon, Nereid, was discovered in 1949; oddly enough, although Nereid revolves around Neptune in a normal direction, Triton circles its planet in a 'retrograde' fashion – in the reverse direction.

The discovery of Neptune had made one thing clear: that Bouvard was mistaken when he wondered if Newton's law of inverse squares would apply to a planet as distant as Uranus. It clearly did. If that was so, then it was quite possible that there could be another planet in the solar system – since the sun's gravity is sufficient to hold a planet at more than four times the distance of Neptune.

Only five years after the discovery of Neptune, astronomers were intrigued to learn of the discovery of yet another unknown planet. What tantalized them was that it had apparently been lost again. James Ferguson, a member of the staff of the US Naval Observatory, had been studying the movements of the asteroid Hygeia, and he had noted its position at a certain time by a familiar method – that is, by observing its relation to various stars in the vicinity. An English astronomer, John Russell Hind, was studying Ferguson's report, and noticed that one of Ferguson's 'stars' was missing. It was on Ferguson's map, but not in the sky. The director of the Naval Observatory checked Hind's observation, and admitted ruefully that the 'star' Ferguson called 'k' was indeed missing. Which meant, almost certainly, that it was not a star but a planet. An unknown planet. Ferguson tried frantically to re-locate his wandering star, but it had moved into Sagittarius, with the Milky Way behind it. The scientist Willy Ley has compared his problem to trying to locate one particular street lamp in a large city as you fly over it in an aeroplane. Ferguson gave up in despair. He may or may not have been consoled when, a few years later, an astronomer named Peters produced evidence that he had made a simple error in using a device called a filar micrometer (for measuring the apparent separation of stars), and that his 'planet' had never existed.

In the year Peters made this suggestion – 1879 – the French astronomer Nicolas Camille Flammarion renewed interest in the search for the unknown planet beyond Neptune when he pointed out that it might be betrayed by its influence on the orbits of passing comets. Many astronomers were already attempting to duplicate the feat of Adams and Leverrier, and discover 'Planet X' by calculating slight irregularities in the motions of Neptune and Uranus. Yet these irregularities were so slight that they might well be caused by other factors. Flammarion's suggestion renewed hope, yet also complicated the problem; for if the planet could influence comets, the comets could also influence the planet. . . .

Two American astronomers joined the search in the twentieth century: William H. Pickering and Percival Lowell. They obtained widely differing results, but inspired many followers. Lowell himself began an intensive search for 'Planet X' in 1905, using the newly developed methods of photographic astronomy. Professor Max Wolf of the Königstuhl Observatory had used photography in his search for asteroids, and – as we have seen – was highly successful, discovering them by the hundred. His method was simply to point his camera at the sky, and leave a plate exposed for a few hours. The fixed stars would stay fixed; the planetoids would show as a line. (Of course, the camera had to be mounted on clockwork to follow the stars as they moved with the earth – otherwise they would have been lines too.) Unfortunately, a planet beyond Neptune would move so slowly that this method would not work. (Neptune takes 165 years to complete a revolution around the sun, and 'Planet X' would take even longer.) What Lowell had to do was to photograph a patch of sky, then photograph the same patch several days later, and carefully compare the two photographs to see if anything had moved. The problem was complicated by variable stars, which change their brightness over a period, and by flaws in the photographic emulsion. After two years, Lowell had to admit failure. He had discovered several new asteroids, and even comets. But no planet.

He began a second search in 1914; but it ended when he died in 1916. In 1919, the Mount Wilson Observatory decided to use Pickering's calculations to try to find 'Planet X'. They were just as unsuccessful as Lowell.

In the 1920s a new invention aided the search. It was called a 'blink comparator'. This works, in effect, by focusing two star-photographs, one on top of the other, then causing them to alternate at great speed, so that first one photograph appears, then the other. Because the retina of the eye retains images for a moment after it has seen them, any 'star' that has changed position between the two photographs appears to 'jump'.

At this point there enters the story another poor boy determined to make good. His

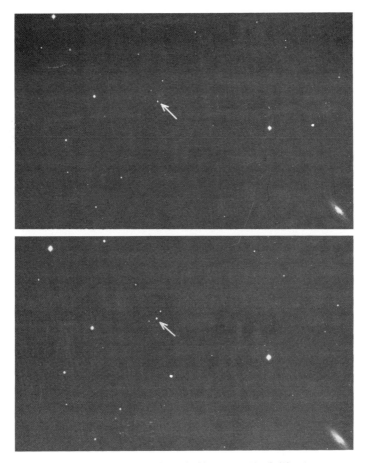

Photographs taken in the Lick Observatory in California
showing the motion of Pluto in one day

name was Clyde William Tombaugh, and he was born in 1906 on an Illinois farm. When Clyde was twelve, an uncle who had a small telescope invited him to look at the moon. What he saw startled Clyde as much as it had startled Galileo three centuries earlier. He began to buy cheap books on astronomy and borrowed everything he could find on the subject in the local library. He built his first telescope the following year, and was equally delighted with his first sight of Mars.

When his father told him they lacked the money to send him to college, Tombaugh decided to continue his studies on his own – while he worked all day on the farm. Like Herschel, he ground his own mirrors for larger telescopes. He was soon convinced that he had seen the 'canals' of Mars. (These markings do not appear to be entirely a matter of wishful thinking; many astronomers have seen them clearly, then found that they had vanished the next night.) In 1928, when he was twenty-two, he sent some of his sketches to the Lowell Observatory, asking their advice. The director, Vesto Melvin Slipher, invited him to come and join the staff. So in 1929, Clyde Tombaugh joined in the search for 'Planet X'.

The summer was pleasant enough in Flagstaff, Arizona; but the winter nights were icy. The dome cannot be heated because heat would make waves in the air, which would cause the stars to ripple. Tombaugh dressed in his thickest clothes, and sat shivering as he peered into the eyepiece.

In January 1930, one year after his arrival at Lowell, Tombaugh took three photographs of the eastern part of Gemini. Then he began the long business of 'blinking' them. A whole plate contains thousands of stars; he had to do it piecemeal. On 18 February, he began blinking a section of the plate near Delta Geminorum. Suddenly, he saw a slight movement. A dot appeared to jump from one place to another.

It *could* be an asteroid; but that was unlikely. The movement was very small, suggesting that it was further away. The 'jump' was a mere three and a half millimetres – too little for an asteroid, which would move much further in the week between the photographs.

Tombaugh had no doubt: he had found 'Planet X' – the planet we now call Pluto. He rushed across the hall to tell his immediate superior, then the two of them went to tell Slipher. He was excited, but warned that they must be cautious. By then it was too cloudy to take another photograph of the region; but the following day was clear, and Tombaugh made a one-hour exposure. The next day, he checked it against one of his original plates. The unknown object had moved a centimetre from its position of a month earlier. There could be no possible doubt.

The discovery of 'Planet X' was announced on 13 March 1930, the seventy-fifth anniversary of Lowell's birth. Unfortunately, Lowell had been dead fourteen years. If the blink comparator had been invented twenty years earlier, he might easily have made the discovery himself.

Yet if Lowell had discovered 'Planet X' he would have had to admit that some of his own speculations were wide of the mark. He had calculated that it ought to be another giant, like Jupiter, Saturn, Uranus and Neptune. Instead, Pluto was the samc size as our moon, with a diameter of two thousand miles. Moreover, it differed from all the other planets in one important respect: its orbit was not in the same plane. Mercury is inclined to this plane at an angle of 7°; Pluto is tilted at 17° to the ecliptic. This unexpectedly high inclination was another major reason why Lowell did not find the planet: naturally, he searched along the plane of the ecliptic. Tombaugh struck lucky at a time when Pluto was at a point in its orbit when it was plunging down towards the ecliptic. Moreover, at its closest point to the sun (its perihelion), it was actually closer than Neptune. So its orbit crosses that of Neptune. To add to the confusion, it is roughly in the orbit predicted by Bode's Law for Neptune.

So why *is* Pluto such an anomaly? One suggestion is that it is not a planet at all – merely an escaped moon of Neptune. Is this why Triton has a retrograde orbit?

Like the story of Uranus and Neptune, the story of Pluto cannot be concluded in this chapter – or this book. In 1978, James Christy and Robert Harrington of the US Naval Observatory discovered that Pluto has a small moon of its own, Charon, but that it must be five times smaller and forty times lighter than was believed earlier. This makes it a great deal smaller than our own moon. The latest speculation is that Pluto is merely a 'snowball' of frozen gas which may have drifted into our solar system from somewhere beyond – perhaps from Öpik's 'comet reservoir' – or else some other piece of 'space debris'. Which suggests that the unknown planet beyond Neptune might yet be discovered.

As our demon turns his space capsule back towards the earth, I suspect I hear him mutter irritably: 'And what *difference* will it make?' I have to admit that he has a point.

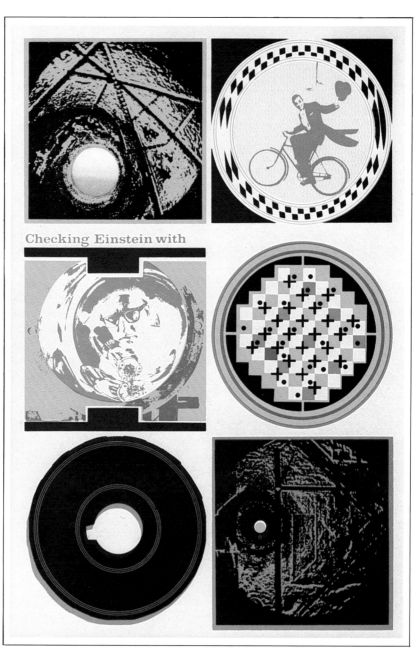

Modern abstract; print from Eduardo Paolozzi's Moonstrip Empire, Vol. I

CHAPTER EIGHT

THE WORLD TURNED INSIDE OUT

Our story now must move from the depths of outer space and back into the physics laboratory. For it is impossible to understand the realm of the stars without understanding the inside of the atom.

In fact, no one even realized that the atom *had* an 'inside' before the end of the nineteenth century. The discovery involved the apparatus known as a cathode ray tube – by means of which Roentgen discovered X-rays in 1895. This is simply a glass tube containing a vacuum, with an electric terminal at either end. When Sir William Crookes first succeeded in sucking most of the air out of such a tube with a pump, he discovered that the remaining air began to glow when he switched on the current. He had discovered the principle of the neon light. But when more air was pumped from the tube, the glow gave way to a dark space around the cathode (the plate emitting the current) which expanded, appearing to push the glow down the tube until only the far end was luminous. It looked as if the electricity was turning into some sort of ray, and streaming down the tube. (Early science fiction writers were quick to take up the idea of death rays.) Crookes discovered that, if focused, these rays could melt metal.

But what were the rays? If a magnet was held near the tube, they could be deflected from their course; so it looked as if they were charged particles. That was discovered as early as 1858. But it was not until 1897 that the English physicist Joseph John Thomson showed that the rays could also be 'bent' by passing them near metal plates charged with electricity. And since he knew how much electricity he had put into the plates, he was able to work out the size of the charged particles. To his astonishment, they proved to be far smaller than a hydrogen atom – the lightest atom known to science. These tiny particles were called electrons.

It began to look as if Newton was right after all. If electricity consists of tiny charged particles, then perhaps light does too? But this would contradict Maxwell's theory that light, heat and magnetism are all different kinds of 'waves'. The evidence for Maxwell's theory was now very powerful. Roentgen had discovered that a cathode ray tube also produces X-rays, which have a shorter wave length than ordinary light, and Heinrich Hertz had produced radio waves, which have a far longer wave length. He did this by simply allowing an electric spark to jump back and forth between two metal balls. He detected the waves by means of an ordinary piece of wire bent into a circle, with a tiny gap between the ends. Although this wire was in no way connected to the

The discoverer of X-rays, Professor Wilhelm Roentgen, in his laboratory

two metal balls, sparks nevertheless began leaping across the gap. So the metal balls must be producing some form of electrical wave, which was being picked up by the wire loop – just as a singer can make a glass on the other side of the room vibrate by singing a certain note. The wire loop became known as an 'aerial'.

In 1895, a youthful Italian inventor named Guglielmo Marconi realized that if an electrical machine can send out 'long waves' and an aerial can pick them up, then there is no reason why electric power should not be transmitted without wires. He only had to combine this knowledge with Alexander Graham Bell's discovery of how to turn electric current into sound, to produce the invention known as the 'wireless'.

Now an advocate of Newton's particle theory of light might argue that X-rays are simply very fast particles which, because of their speed, can penetrate matter. But in that case, radio waves would have to be very slow particles. And it would be difficult to explain how they can pass through the walls of buildings. So, on the whole, Maxwell's wave theory seemed to have won the day.

Or almost. There *was* one embarrassing little anomaly, which became known rather melodramatically as 'the ultra-violet catastrophe'.

Think of the whole range of electromagnetic waves – which includes light – as an enormous piano keyboard, with radio waves down at the bass end. How far does the keyboard extend in the other direction – up beyond ultra-violet light and X-rays and gamma rays? There is no obvious limit; it could go on for miles.

But there is an obvious difference between a piano keyboard and Maxwell's waves.

The earliest form of mobile radio, 1901. Marconi stands at the extreme right,
next to Ambrose Fleming, inventor of the thermionic diode

Each note on the keyboard is separate. In Maxwell's wave theory, all the notes are joined together – radio waves blend into heat, then into light, then into X-rays, and so on. So in theory, if you strike any note, it ought to make all the others vibrate. And when you switch on your electric fire or cooker, it ought to send out X-rays and gamma rays – rot to mention violet and ultra-violet light. This is the so-called 'ultra-violet catastrophe', which fortunately, is purely theoretical. But why is it theoretical? Why are the 'notes' on Maxwell's piano divided from one another?

In 1900, a Berlin physicist named Max Karl Ernst Ludwig Planck produced an interesting explanation. Suppose energy is divided into small, individual packages. This would mean that it is not free to flow in all directions – up and down the 'keyboard' – but would be much more restricted. Think, for example, of what happens when you empty a cup of water over a table: it runs freely all over the surface. But if the water is converted into powdered ice, you can empty it on the tabletop, and it remains in a heap.

The illustration is, of course, crudely oversimplified; but it conveys the spirit of Planck's 'quantum theory'. He was saying, in effect, that light is both particles *and* waves. It consists of waves tied up in small packets. The packets can come in different

sizes, depending on whether one is talking about radio waves or X-rays; but their basic unit is always the same. (It is known as 'Planck's constant'.)

Most of Planck's fellow scientists found the idea untenable. Why *should* energy come in packets? After considering their objections, Planck was inclined to agree with them. It *did* seem slightly arbitrary. After all, one can imagine a particle flying through space, or a wave moving through water, or the 'ether'. But what does a 'parcel' of waves look like? Planck tried a compromise, suggesting that light *travels* in the form of unbroken waves, but is emitted – or absorbed – in quanta. But this seemed to raise more problems than it solved. Planck himself began to lose faith in his theory.

In the Swiss Patent Office in Berne, a twenty-four-year-old clerk was more convinced by Planck's quanta than Planck himself. His name was Albert Einstein, and he was working in the Patent Office while trying to find a teaching job. He had always been fascinated by the problem of light. At sixteen he had asked himself the question: what would it be like if one could travel as fast as a beam of light? The answer seemed to be that it would not be 'light' any more – just as a wave on the sea would not be a wave if you could travel along smoothly on the crest. A man travelling at the speed of light would see light as a *static* electrical vibration. And that is a contradiction in terms. . . .

He was also fascinated by an anomaly that had been uncovered by Hertz's assistant, Philipp Lenard. In his experiments with radio waves, Hertz had accidentally discovered the 'photoelectric effect' – when light falls on a sheet of metal, it 'evaporates' electrons from its surface, just as sunlight evaporates molecules from a sheet of water. But according to Maxwell's equations, the speed of the 'evaporated' electrons ought to increase if the light became more intense – that is, if the experimenter used a 150 watt bulb instead of 60 watts. And this did not happen. What *did* happen, Lenard discovered, was that the *number* of electrons increased. On the other hand, if the wave length of the light was changed – for example, if violet light was used instead of yellow – the speed of the electrons increased. This contradicted Maxwell's predictions.

In 1905, Einstein sent a paper to *Annalen der Physik* (*Annals of Physics* – the German counterpart of *Nature*) in which he pointed out that Planck's theory would explain the anomaly. If light is made up of 'packets' of energy – like explosive shells – and blue light contains more explosive than red, then you would expect blue light to send the electrons whizzing off at a greater speed. If you merely increase the number of red 'shells', then they will knock out more electrons, but will do it less violently. So, according to Einstein, light actually travelled in packets. Planck himself found the idea difficult to accept, and it was several years before he realized that Einstein was right. Light behaves like waves *and* particles.

It was Einstein's obsessive preoccupation with the problem of light that led him to the next major step, which occurred later in the same year, 1905. Einstein said later that from the time the idea occurred to him to the time he completed his paper, only six weeks elapsed. The idea was the Special Theory of Relativity – perhaps the most revolutionary advance in science since Copernicus's book on the solar system.

But before we go on to discuss relativity, it is important to realize that it was not the result of pure genius and inspiration. In his history of electricity, Sir Edmund Whittaker actually ignores Einstein's part in creating Special Relativity, and refers to the 'relativity theory of Poincaré and Lorentz'. He was evidently so exasperated by the

Professor Albert Einstein

view of Einstein as a super-genius that he went to the other extreme. When the book appeared,[1] reviewers berated Whittaker for wild eccentricity. But there is much to be said for his stand. Relativity was not merely 'in the air'; it was actually on paper and in print by 1905. Jules Henri Poincaré had outlined the theory in 1904 at a science congress in St Louis, and had stated Einstein's famous conclusion: 'From all these results there must arise an entirely new kind of dynamics, *which will be characterized above all by the rule, that no velocity can exceed the velocity of light.*'[2] This was published in 1904, a full year before the idea occurred to Einstein.

In fact, the basic problems that led to relativity had been recognized for well over a quarter of a century before Poincaré and Einstein grappled with them head-on. They had been around since the time of Galileo. We may recall that Aristotle stated that if an object is dropped from the top of the mast of a moving ship, it will fall behind the mast; in other words, the body drops down in a straight line, but the ship moves on. This is untrue. For the object has been given a certain *forward* motion by the ship, just as if it had been thrown by hand. So (provided there is no wind) it would still land on the deck at the foot of the mast. This was one of the many errors of Aristotle that Galileo exposed. Galileo knew that all laws of nature are the same, whether tested on dry land or at sea (provided the ship was moving smoothly and uniformly).

Let us suppose that a child has been given a 'physics set' for Christmas, containing various pieces of apparatus with which he can perform experiments in light, heat and mechanics, and that he decides to perform some of these experiments on a jet aeroplane crossing the Atlantic at a speed greater than sound. Assuming that the aeroplane is flying smoothly, is there any way in which he can tell that he is moving – without looking at the flying clouds out of the window? The answer is no. Every experiment he performs will come out exactly the same as if he'd performed it on the ground. And this, after all, is reasonable. For in a sense, we are *always* in a jet aeroplane; our earth is spinning at a thousand miles an hour, as well as rushing headlong around its orbit at 65 000 miles an hour, and (together with the rest of our solar system) in the direction of the constellation Hercules. If motion through space affected experiments, we would have a highly distorted view of the laws of nature.

This statement seems perfectly reasonable and uncontroversial until we recollect that, according to Maxwell, all electromagnetic waves – light, radio waves, X-rays – move at 186 000 miles per second. (We have to add: in a vacuum. Light travels more slowly in water or air.) That sounds very much as if Maxwell is saying: there is no way of making light go faster. And if that is so, the speed of light is a *law of nature*.

But surely, there *is* a way of making light go faster? If you are on a train, and you run along the corridor, your total speed is the speed of the engine plus your own. (Or, if you are running towards the rear of the train, the speed of the train minus your own.) Similarly, if you shoot a stone from a catapult along the train corridor, its speed would be its own speed plus the speed of the train. The same applies if you fired a bullet from a gun. So if you switch on an electric torch and point it towards the engine, the speed of the light *ought* to be its own speed plus that of the train.

But if that were so, light would have broken Maxwell's law of nature, which says that electromagnetic vibrations cannot go faster than 186 000 miles a second. How can we resolve this contradiction? If Poincaré is right, we must begin from the assumption that under no circumstances can light travel faster than 186 000 miles a second.

Well, if the law of nature will not 'bend', then something has to. What is it? Suppose, for example, our train was travelling at half the speed of light. Then surely it would be quite impossible for the light from the torch *not* to travel faster than light? How would it wriggle out of this dilemma? What would 'bend'?

Einstein's answer was breath-taking. It was: space and time.

Let us go back for a moment to the man dropping a weight from the mast of the moving ship. From his point of view, the weight has only dropped the height of the mast. But if there is a man standing on the jetty, watching the ship go past, he can see that the ship is moving too; so the path of the weight is not perpendicular, but a diagonal. From his point of view, the weight has travelled further. Who is right? Common sense says the man on the jetty. Einstein replies: no. From the point of view of the universe, both of them are equally right. He calls the ship and the jetty 'reference systems'. This means that, if you wanted to make really accurate measurements, you could mark them off in inches or centimetres – up, along and across – and specify exactly how much further the weight had travelled according to the man on the quay. Einstein asserts that, where the laws of physics are concerned, one reference system is as good as another.

Or think of two men sitting on either side of a railway carriage, tossing a ball to one another. From their point of view, the ball moves in a straight line from one side of the carriage to the other. But if you happen to be standing on a station platform as the train flashes past, you will see that the path of the ball is a V; between the time the ball has left the hand of the man on the left, and the time it has returned to his hand, the train has also travelled several yards.

But if the path of the ball is a V and not a straight line, the ball has travelled *further* from your point of view on the platform. And if you on the platform and the men in the train try to measure the speed of the ball, you will get different results. But how can the same ball be travelling at two different speeds at the same time?

There is only one way to make the sum come out right, says Einstein. All we have to do – and you should take a deep breath before you try to swallow this one – is to recognize that *time in the carriage goes slower*.

And now we come to the most bewildering part of the argument. If Einstein is right, and one reference system is as good as another, then it doesn't matter whether we say that the train is moving past the station or the station is moving past the train. In which case, the men on the train will believe that time on the station platform goes slower too. Suppose now that it is two porters who are throwing the ball. The same argument applies; the ball appears to travel in a V from the point of view of the men on the train.

Another concept that Einstein questions is the notion of things happening *simultaneously*. To explain this, I will borrow an illustration from Bertrand Russell. Let us suppose that brigands are standing on the station platform, and as the train goes past, they simultaneously shoot the engine driver and the guard. An important legal case hangs on the question: did they die simultaneously, or did one die a moment before the other – a rich uncle has left all his money to the family of whichever died first. A guard standing midway along the platform testifies that the men died simultaneously, because he heard both shots at exactly the same moment. But an old gentleman halfway along the train argues that the engine driver died first, because he heard that shot a moment earlier than the shot from the rear.

The judge points out that the old gentleman has made an honest mistake. Because the train is travelling towards the brigand who shot the driver, the sound takes less time to reach him than the shot from the back of the train – whose sound has to catch him up. So the family of the engine driver gets the cash.

This, Einstein would say, is understandable, since the judge is sitting in a courtroom on earth. But if, instead of a train and a station, we imagine two long space ships somewhere in the upper atmosphere, with the brigands (and station master) on one of them, and the driver and guard on the other, we can see that it is far more difficult to determine whether the shots were simultaneous, since we are now no longer sure which space ship was moving and which was standing still. From the strictly scientific point of view, one reference system is as good as another. You cannot say that two events were *absolutely* simultaneous – only that they are simultaneous for one reference system (the platform) and not another (the train).

There is another important consequence of this argument. If you are on the platform, and the train is travelling very fast – say half the speed of light – you would see the train as *shorter* than if it was standing still. The odd thing is that the men on the train would also see the platform as shorter. How is that possible? This, in fact, is fairly easy to understand, for it comes within our common experience. If you are sitting on the platform when a train rushes through, you have no sooner seen the engine than you are looking at the back of the last carriage; it seems shorter. And for the people on the train, the same thing applies to the platform. It also seems shorter. The effect is reciprocal – like the apparent slowing down of time.

Shorter by how much? It was when Einstein came down to working out the precise mathematics that he discovered an interesting thing. The formula had already been worked out by somebody else – a Dutch physicist named Hendrik Antoon Lorentz. But Lorentz had worked it out in connection with a quite different problem. It was the problem raised by the famous Michelson-Morley experiment.

Before this can be described, we need to know, first of all, that scientists had been brooding on this question of 'reference systems' ever since 1818. In their experiments concerning light, they had recognized that they needed some kind of solid foundation. The earth is moving, the solar system is moving – so how could they be sure that this movement did not affect the light in their experiments? It would be convenient, for example, if science could point to the exact centre of the universe, and say, 'When I talk about motion, I mean *absolute* motion in relation to that.' But the centre of the universe was obviously too far off to be of any help.* A mathematician named Carl Neumann tackled the problem in 1869, and suggested that this 'absolute reference system [if he had been a theologian, he might have felt that the main railway station in heaven would meet the case] should be labelled "the body alpha"'.[3] Other mathematicians suggested that the body alpha should be the 'fixed stars' – this was before it was generally realized that they are not fixed.

The answer, someone suggested, was surely the 'ether' – the invisible jelly in which light travels, and whose waves are electromagnetic vibrations. If that permeates the whole universe, then it has to be standing still: there is nowhere for it to move to. In 1887, two American scientists, Edward Williams Morley and Albert Abraham Michelson, devised an ingenious experiment to find out just how fast the earth is travelling in relation to the ether. As the earth moves through the ether, the ether also moves past the earth, causing an 'ether wind'. What they proposed to do was to shoot a beam of light up and down the 'ether wind', and another beam across and back, and see if – or rather, by how much – the journey across was shorter than the journey up and down.

We can understand the principle of the experiment if we imagine a swimmer

---

* Many scientists would deny that the universe has a 'centre', claiming that this is a popular misconception based on trying to visualize an n-dimensional universe in three-dimensional terms.

Time is out of joint. The astronaut confronted with the old man.
Closing scenes from the film 2001, A Space Odyssey *by Stanley Kubrick*

swimming across a river and back, and his twin brother swimming up the river and down at the same rate. A little simple mathematics shows that the swimmer who goes across the current and back will take less time than the one who goes up and down. How much faster will depend on the speed of the current. So to find the speed of the 'ether wind', Michelson and Morley only had to direct their two beams of light at right angles, having them reflected back by mirrors. The light travelling in the same direction as the earth should take longer; *how much* longer would reveal the speed of the 'ether wind'.

Both beams of light took exactly the same time. How was that possible? In 1895, the Irish physicist George Francis FitzGerald suggested an explanation. If objects are made up of tiny electrical particles, then you might expect them to contract as they fly through the air – just as a spear made of some compressible material would contract as it left the thrower's hand. So perhaps Michelson and Morley's measuring rods had also contracted in the direction of the earth's motion, and therefore the light travelling up- and downstream *seemed* to have travelled further than it really had?

Lorentz worked out mathematically just how much this contraction would have to be in order to cancel out the result of the experiment. His formula was called 'the Lorentz transformation'.

It was this formula that Einstein arrived at when he tried to work out how much the train would appear to 'contract' as it flies through the station. But there was a basic difference between Einstein's approach and Lorentz's. Lorentz thought the contraction was real, caused by the squashing together of atoms. Einstein did not think it real. The train only appears to contract, from the point of view of people on the platform. (More precisely, the train would turn red – because the wave length of the light is 'stretched' as it travels away from you – and it would appear to rotate slightly away from you, so you would see it foreshortened.)

There is another curious consequence of Einstein's theory, which was again anticipated by FitzGerald. The Irish physicist worked out that if a twelve-inch ruler could fly through the air at 161 000 miles per second, it would contract to six inches. Einstein's formula said the same. But it also meant that if the ruler travelled at the speed of light, it would contract away to nothing. However, one of the consequences of Einstein's theory had already been stated by Poincaré – that nothing can exceed the speed of light. Why? Because its mass increases as it goes faster. In 1882, Thomson had noted that if an object was electrified when it was in motion, then according to Maxwell's equations, it would become heavier; and its mass would increase with speed. (We need to note, incidentally, that weight is not the same thing as mass. A man would be far lighter on the moon, but his *mass* would remain the same as on earth.) Einstein's equations verified this, and showed that as the speed approaches that of light, the mass gets closer to infinity.

And this, of course, was the whole aim of Einstein's Theory of Relativity – to work out what the consequences would be if nothing could travel faster than light. As we have seen, those consequences are very strange indeed. Perhaps the strangest is the thought that time is 'changeable' for different observers. Einstein gave a dramatic illustration of this, which has become known as the 'twins experiment'. One of two twins sets out in a space ship at half the speed of light to make a journey to the nearest star; the other brother stays at home. Ten years later, he returns. His twin on earth is visibly older, with a few grey hairs; but the space traveller looks nearly as young as on the day he left; time has almost stood still for him because of his speed. If his earth-bound brother could have communicated with him by means of a television set while he was in flight, he would have wondered why the space traveller was moving incredibly slowly, why it took him a whole minute to finish a short sentence. The answer would be that his 'time' had slowed down; that the atoms in his body were moving at a fraction of their normal speed.

Would this really happen? There is a considerable amount of evidence that it would. There are certain sub-atomic particles that can exist for only a fraction of a second; physicists have accelerated these to speeds approaching that of light, and the particles have lasted longer, just as predicted.

Naturally, these results have little or no consequence for human beings, since our speeds are usually so small. Practically speaking, it means that if you are on a train, and you run along the corridor towards the engine, your 'total speed' is *not* your own speed plus that of the train, but slightly less. For those of mathematical inclination, I will give the Lorentz formulae. The actual distance you travel, x, will not be $x_1 + x_2$,

Einstein's laboratory at Zurich's Technische Hochschule

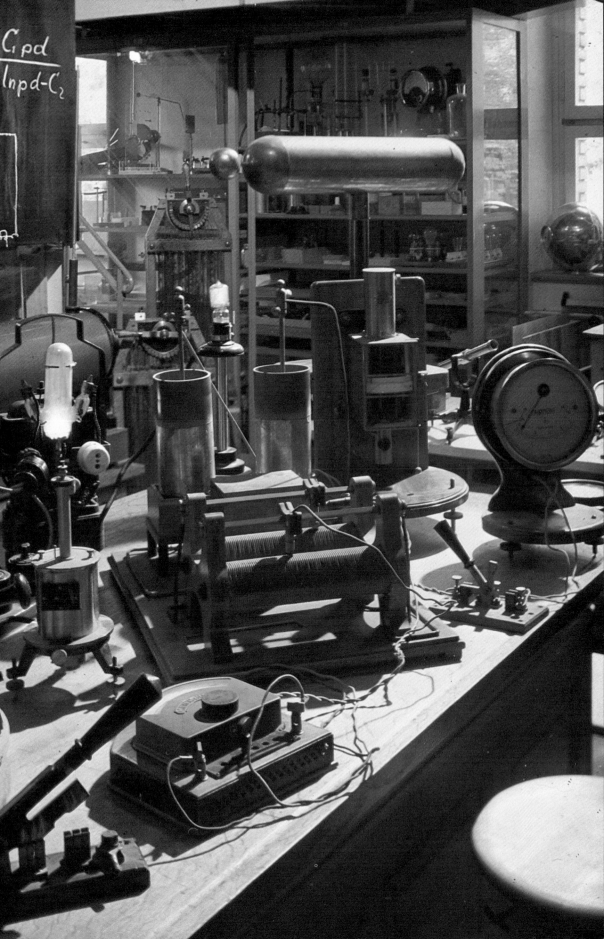

where x_1 is the distance you travel, and x_2 is the distance travelled by the train, but this same distance multiplied by the following formula:

$$\sqrt{1 - \frac{v^2}{c^2}}$$

where v is the 'total speed' (of train plus man) and c is the speed of light. It can be seen that, compared to 186 000, v is going to be a very tiny figure indeed, so that the fraction

$$\frac{v^2}{c^2}$$

is going to be very small; so the formula amounts to the square root of one, which is one. But for someone travelling at 161 000 miles per second,

$$\frac{v^2}{c^2}$$

becomes three-quarters, and the formula becomes equal to a half. The total distance is halved. If we want to work out the speed, we cannot follow the usual system of dividing this distance by time, for (if we recall), time is also altered by speed. The new time formula is:

$$\frac{t - \frac{v}{c^2} \times x}{\sqrt{1 - \frac{v^2}{c^2}}}$$

Here again, x is the 'total distance'. Again, this becomes lower and lower at high speeds, so that the distance covered becomes steadily smaller as we approach the speed of light. Why? Because, according to Einstein's formula, the mass increases, so that it costs more and more energy to accelerate.

These arguments of Einstein, presented in his paper 'On the Electrodynamics of Moving Bodies' in 1905, made him famous in the scientific community, even if the majority of scientists felt there was a catch somewhere.– rather as in Zeno's story of Apollo and the Tortoise, which 'demonstrates' that Apollo can never overtake the tortoise, no matter how fast he runs. (I must admit that I *still* feel this whenever I run over the basic arguments for relativity.) His theory did not bring instant fame; in fact, in 1907 he was turned down for a post of *privatdozent* (unpaid lecturer) at the University of Berne. The decision was reversed a year later.

His achievement had been considerable for a man still under thirty; but he was unable to relax. A further problem of relativity continued to bother him. The Special Theory states that if you are travelling in a train in a straight line at a regular speed, you cannot detect that you are in motion (this is an ideal train without bumps or jerks) unless you look out of the window. But if the train does not travel in a straight line – if it goes round a bend – or if it accelerates suddenly or slows down, you *can* tell. Einstein wanted to find a way of applying relativity to this problem too. And acceleration brought to mind the problem of gravity, since falling objects go faster by the second. It occurred to Einstein that when a train goes round a bend, pulling you sideways in your seat, it is just as if a force of gravity has pulled you sideways. So he began to see the problem of extending Special Relativity to *all* motion as a problem involving

Albert Einstein explains his theory of relativity to British Prime Minister
Ramsey MacDonald in Berlin, 1931. On the left is Max Planck, whose quantum
theory is essential to nuclear science

gravity. Gravity *behaves* as if you were going round a bend in a train. And, using the same approach as in the Special Theory, he asked himself whether, for practical purposes, they could be considered the same thing. That is, whether acceleration or a change in direction can be looked upon as a kind of gravity. He used a famous illustration: if a man in outer space was enclosed in a lift, and the lift began travelling upwards – dragged by some demon – at an increasing speed, the floor would press up against his feet, and produce *exactly* the same effect as gravity. Einstein called this insight 'the principle of equivalence' – the principle that gravity and acceleration are, for practical purposes, the same thing. He enunciated this principle in 1908.

The earth, however, is *not* being accelerated by a space demon. And if it were, gravity would only apply to its top half: the rest of the human race would fall off the bottom. So how can the principle of equivalence be applied to our universe? In formulating his reply, Einstein may have been inspired by a remark made by Fitz-Gerald. He said, 'Gravity is probably due to a change of structure of the ether, produced by the presence of matter.'[4] This can only mean that matter causes some kind of distortion in the ether – that hypothetical jelly whose waves are called light – which gives the impression of a force – just as going round a corner gives an impression of gravity. Once again, we could object that our earth is a globe, and a force that made things press against one side would cause them to fall off the other.

Einstein also had a tentative explanation for this problem. In the 1880s, an ingenious writer named Edwin Abbott wrote a book called *Flatland* about a two-dimensional world – like a sheet of paper – where the inhabitants (squares, circles, and so on) could not imagine the meaning of the word 'height'. Abbott was trying to challenge his readers to imagine that our universe might have a fourth dimension – in addition to height, length and breadth. (It was a concept that had been familiar in mathematics since the late eighteenth century.) Around the turn of the century, a writer named C. H. Hinton accepted the challenge, and tried to work out the consequences of a world with four dimensions. But the strange distortions of our normal concepts by Special Relativity suggested another possibility: that *time itself* is the missing 'fourth dimension'. If you agree to meet someone on the corner of Regent Street and Oxford Street, you have specified two dimensions – length and breadth. If you specify that you will meet them on the third floor of a building there, you have specified the third dimension. But these explanations will be useless unless you add what *time* you will be there; that is the fourth dimension.

We know that our three dimensions of space are really inseparable – Abbott's flatland could not exist. And the Special Theory has shown that space and time are inseparable – both vary with speed. One of Einstein's old professors, Hermann Minkowski, had actually constructed a mathematics of four dimensions. Einstein now took advantage of this to formulate his new theory of gravitation. Instead of space *and* time, said Einstein, we must try to think in terms of a single undivided entity called space-time. Imagine a being who possesses *five* dimensions, looking down on our space-time universe just as we can look down on a sheet of paper. He would not see you and me as three-dimensional creatures travelling through time; he would see us as four-dimensional beings, a kind of solid line with a foetus at one end and a corpse at the other. We do not have this advantage; we can only see three-dimensional slices of this four-dimensional being.

According to Einstein, our five-dimensional being would not see a force of gravity holding us down to the earth. After all, there is something very odd about gravity; it is supposed to act on distant objects with only empty space between them (Newton called it 'action at a distance'), and it attracts heavy objects at the same speed as light ones. The five-dimensional man would see something far less paradoxical, something more like the way that a river valley causes water to flow along it. Einstein argues – like FitzGerald – that matter causes a distortion or warp in the 'ether', or in space-time. You can think of space-time as being something like a sheet of rubber, in which a heavy object, like a star, makes a bulge or hollow. And if a smaller object rolls across the sheet and goes near the star, it will naturally begin to roll down into the concavity. This is gravity. The object does its best to travel in a straight line and, in a sense, it succeeds – just as the old Roman roads went up hill and down dale in order to avoid going around corners. Our earth is caught in the 'bulge' created by the sun (although, of course, it also creates its own small dent), and it runs round and round the surface of the 'hollow' just as a roulette ball runs around the circumference of a roulette wheel.

Einstein began to publish the results of his reflections on 'General Relativity' in 1915. As with the Special Theory, the new mathematics of four dimensions gave more or less the same results as classical mathematical physics. But, as with the Lorentz transformation, there were slight variations when dealing with large figures. And it was one of these slight variations that suggested that Einstein's four-dimensional theory was more accurate than the mechanics of Newton's *Principia*. Newton's formula fails to explain why the perihelion of Mercury moves around by a few seconds of an arc every century (the precise figure is 43 seconds), causing the whole orbit of the planet slowly to rotate. Einstein's slightly different formula predicted precisely such a movement. Einstein's announcement of this interesting discovery, just before he published the General Theory, predisposed the scientific community to accept the new ideas. The result was that four years later, after the end of the First World War, this community decided to subject General Relativity to a crucial test. For another of Einstein's assertions was that light should be subject to the same laws as matter, and should be affected by a powerful gravitational field. In effect, he was saying that a body like the sun is surrounded by a 'warp', rather like the distortions in an old-fashioned window-pane. So light passing close to the sun should be deflected by it.

This prediction was difficult to verify because the sun's own light is so dazzling that we cannot see whether the light of a star is 'bent' by it. The only time this could be done

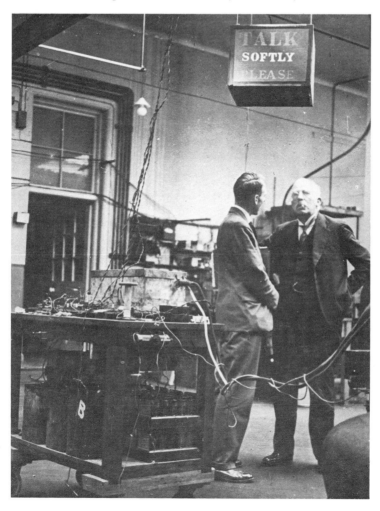

Experimental physicist Ernest Rutherford (right) in the Cavendish Laboratory, Cambridge

was during an eclipse of the sun; and such an eclipse was about to take place in May 1919. A committee headed by the British astronomer Arthur Stanley Eddington (later Sir Arthur) prepared to test Einstein's theory by sending expeditions to observe the eclipse in Brazil and on Principe, a small island off Nigeria. Eddington himself went to Principe, and took the crucial photographs on 29 May. The aim was to see whether certain stars that were 'close to' the sun (as seen from earth) would apparently be displaced. They were. The result was announced six months later, at a meeting of the Royal Astronomical Society. The sense of drama was tremendous, since other astronomers had announced tentative results that seemed to show that Einstein's figures were wrong; some made the deflection smaller, some greater. But Eddington's highly accurate readings showed that Einstein was correct within a tiny percentage. The result, as one of Einstein's biographers remarks, is that he awoke on the morning of 7 November 1919 to find himself famous. For the remaining thirty-five years of his life he was a living legend, one of the half-dozen most famous men of his time.

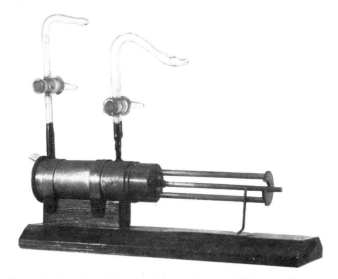

Rutherford's nuclear disintegration chamber. In 1919 he was first to disintegrate a nucleus artificially

Perhaps the most triumphant confirmation of his theory – and the most terrifying – was the one that occurred two decades later, by which time Einstein was living in exile in America. The Italian physicist Enrico Fermi bombarded uranium with neutrons, in the first 'atomic pile', and his results seemed to show that some artificial elements had been created. Otto Hahn and Fritz Strassman repeated these experiments, and confirmed that what had happened was that the nucleus of the uranium atom had been broken apart, to form the lighter element barium. But a barium atom is much lighter than a uranium atom. What had happened to the extra weight? The answer worked out by Hahn and Strassman was: it had turned into energy. Another of Einstein's formulae, $E = mc^2$ (energy equals mass times the speed of light squared), had been confirmed. The missing mass of the uranium atom had escaped as an 'atomic explosion'. The dropping of the first atom bomb on Hiroshima in 1945 brought home Einstein's triumph even more dramatically than the eclipse of 1919.

Like Laplace, Einstein believed that his universe of spherical space-time was stable and unchanging. The Einsteinian universe could be pictured as an enormous balloon whose *surface* is space as we know it. His calculations seemed to indicate that the balloon would neither expand nor contract. However, as early as 1917 the Dutch astronomer Willem de Sitter had modified Einstein's equation for the geometry of space-time and had suggested that the most distant celestial objects would appear to be receding from us. Arthur Eddington – one of Einstein's warmest supporters – was particularly interested in this problem, and in the actual size of the balloon. In the early 1930s, one of his former students, Georges Edouard Lemaître, sent him a paper in which he calculated that Einstein's universe was fundamentally unstable, and that if its equilibrium were disturbed, it would either contract or expand. This was an interesting assertion, because in 1929 the astronomer Edwin Powell Hubble had announced his observation that our universe is expanding, and that the galaxies are receding from us at a speed which is proportionate to their distance – so that the furthest away are travelling fastest. As if Einstein's conclusions about the unreality of space and time (as separate entities) were not enough, it now seemed that the universe was swelling like a balloon about to burst. . . .

To understand how Hubble came to reach this conclusion we must take a brief excursion into the history of nebular astronomy.

When Ferdinand Magellan set out on his voyage to the Spice Islands in 1519, he sailed through the straits named after him near the southern tip of South America and noticed two irregular shaped 'clouds' in the southern night sky; they became known as the Magellanic Clouds. Magellan was unaware of it, but he had observed two of the few nebulae (from the Latin for 'mist' or 'vapour') that are visible to the naked eye. In 1612, the German astronomer Simon Marius noticed another in Andromeda. When Newton's friend Halley turned his powerful telescope on the heavens, he discovered dozens more. His own theory was that they were clouds of cosmic gas or matter, shining by their own inner light. In 1750, a brilliant teacher of navigation named Thomas Wright wrote a book suggesting that the nebulae were other star systems like our own Milky Way. It was altogether too advanced for its time and was virtually ignored. But later in the same century William Herschel became fascinated by the puzzle of the nebulae. When he turned his powerful telescope on them, he found that some of them could be seen as clusters of stars. Others remained blotches of light, like irregular clouds of gas, while there were some – which he called 'planetary nebulae' – with mottled discs. But he could not decide whether his last class of 'lens-shaped' nebulae were made of stars or gas. We have seen that Herschel also made the inspired suggestion that the Milky Way is our own star system, and that it is shaped like a wheel or lens. After Herschel's death, an eccentric and wealthy British astronomer, William Parsons, the third Earl of Rosse, decided to build the biggest telescope ever to try to solve the riddle of the nebulae. This monster was 60 feet long, had a 6-foot wide mirror, and cost £12000 to build. It was so large that it could not be swung around freely, but had to be mounted between two huge walls – which meant it could only be turned in an arc north or south. (The earth, of course, swung it around from east to west.) Even this 'Leviathan', as it was called, failed to resolve the lens-shaped nebulae into anything but shining clouds – some with beautiful spiral forms.

At about this time – in the early 1840s – an Austrian scientist, Christian Johann Doppler, theorized that when a source of light recedes from the observer, its waves are 'stretched out' and appear redder, and that when it approaches him, the waves are 'squashed', and so become more blue (or violet). He tested this wave-motion elaborately by having musicians play on a moving railway car, and noting the change in the pitch of the instruments as they moved towards or away from the observer. The 'Doppler effect' was to have far-reaching repercussions for modern astronomy.

Twenty years later, Kirchhoff and Bunsen's invention of the spectroscope had provided astronomers with a new instrument for studying the stars. In 1863, Sir William Huggins directed his telescope at a 'planetary nebula' in Draco, and looked at the spectroscope. It showed only a single bright line – proving conclusively that the planetary nebulae were not clusters of stars, but clouds of gas. He went on to examine another sixty nebulae, and discovered that about one third were clouds of shining gas; many of the rest were clusters of stars. The baffling 'lens-shaped' type were found everywhere except the Milky Way – small, symmetrical objects, which were either elliptical or spiral. Thirty years later, photographic analysis of this third type of nebula revealed that these, too, were composed of stars. Poe, as we have seen, guessed that they were separate 'island universes', like our own Milky Way. But until someone could discover a reliable method of accurately judging the distance of stars, there was no way of resolving this problem. Bessel's method was obviously useless for most stars, which are too far away to show measurable parallax.

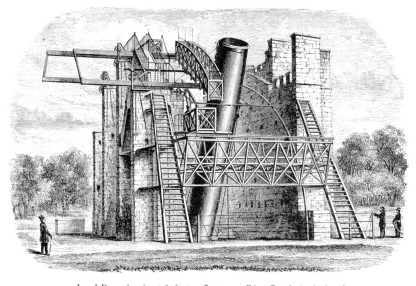

Lord Rosse's giant 6-foot reflector at Birr Castle in Ireland

The answer came through the study of a type of star known as a 'variable'. On 12 November 1782, a deaf-mute, John Goodricke, was examining a star named Algol in the constellation Perseus, and was puzzled to find that it was far less bright than it should have been. Ptolemy's star catalogue had listed Algol (in Arabic it means 'the ghoul' or 'demon') as a star of the second magnitude – that is, only one degree less bright than the very brightest. To Goodricke that night it was plainly a star of the fourth magnitude. Yet the following night it was back to second magnitude. A month later, it was back to fourth again. Goodricke's guess was that Algol is a double star – like Sirius – and that it is periodically eclipsed by its companion. But in 1784, he encountered a more puzzling type of variable. The star Delta in the constellation Cepheus took about four days to become dim, then only one day to brighten. Clearly, it could not be eclipsed by an invisible companion, or the periods would be the same. Astronomers became interested in variable stars, and by the end of the 1880s, 225 of them were listed.*

In 1905, an American astronomer named Henrietta Leavitt, of Harvard College Observatory, began to study the smaller of the two Magellanic Clouds. She was

---

* There seem to be two reasons for the variations. In the case of very large stars, the force of gravity is unable to control the immense bulk, which alternately expands and contracts. In the others, a layer of gas about halfway between the core and the surface is ionized by the tremendous temperatures and expands; at a certain point the expansion causes the temperature to drop; the gas is de-ionized and contracts.

interested in its Cepheid variable stars – those which changed brightness like Delta Cephei – for it was a fair assumption that these Magellanic variables are all roughly the same distance from our earth, whereas variables in our own galaxy might be any distance. At this point, of course, she had no way of knowing that the Magellanic Clouds are not in our own galaxy. But at least she could see that they formed two separate star systems.

She soon noticed an interesting thing about the Cepheid variables: the brighter the star, the longer its period of variation. A variable with a thirty-day period was six times as bright as one with a three-day period. An exciting idea came to her: suppose this applies to Cepheids anywhere in the universe? If it does, then we have a reliable way of judging their relative distances. If two Cepheids appear to have the same brightness, but one has a thirty-day period and one has a three-day period, then we can assume that the former is thirty-six times as far away.

Now what was needed was some approximation of the distance of a Cepheid. This was first achieved in 1913 by the Danish astronomer Ejnar Hertzsprung. He made use of the discovery made by Sir William Herschel – that our solar system is speeding through space at twelve miles per second in the direction of the constellation Hercules. This means that in one year, our sun moves 370 million miles, which is about twice the distance the earth travels from one side of the sun to the other. Hertzsprung chose thirteen Cepheids in our own galaxy, worked out the parallax of each of them to the best of his ability – his accuracy was not all that could be desired, considering their distance and the smallness of the angles involved – then took an average of all thirteen. His resulting estimate of the distance of the Magellanic Clouds – about twenty-six thousand light-years – made astronomers raise their eyebrows; it seemed much too far. (In fact, the Clouds are about eight times that distance.) Moreover, if Hertzsprung was correct, then the Clouds probably lay outside our galaxy; and if they did, the same might well apply to the other nebulae. And, on the whole, astronomers were still inclined to resist this idea. Views had changed little since the beginning of the century when, at an international conference on astronomy, there had been bitter controversy on the question of whether other galaxies existed outside our own. The main champion of the affirmative view was Richard Proctor, but the general feeling seemed to be that he was indulging in wishful thinking – as in his theories on the Pyramid – and the sceptical view prevailed.

In any case, as the size of our own galaxy was still unknown, it was impossible to determine whether the Magellanic Clouds were inside it or not.

And then, in 1918, the American astronomer Harlow Shapley, who was working on globular clusters at the Mount Wilson Observatory in California, repeated Hertzsprung's observation, and came up with more accurate figures. He was also working on the problem first investigated by Sir William Herschel – that of the actual shape of our galaxy. A Dutch astronomer, Jacobus Cornelis Kapteyn, had been working on the same problem since 1906, studying photographs of the Milky Way to find where it was densest. Kapteyn reasoned that if we *are* at the centre of the Milky Way, then the stars around us ought to look equally 'thick'. And this was the conclusion he actually reached in the last year of his life, 1922. Kapteyn's estimate of the size of our galaxy multiplied Herschel's figures by five – he thought it was about fifty-five thousand light-years in diameter. Shapley disagreed. He took into account the uneven distribution of globular clusters. This could be because the globular clusters are arranged symmetrically around our galaxy, and we are out on its edge.

From his calculations of the distance of Cepheids, Shapley worked out that the centre of our lens-shaped galaxy lies about fifty thousand light-years from our solar system, and that we are about three-fifths of the distance between this central point and the edge. More recent calculations have reduced the distance from the sun to the centre to about thirty thousand light-years, and estimated the total radius of our galaxy as being a hundred thousand light-years. Shapley's work encountered much opposition – no doubt partly because he was delivering another blow to human self-esteem. Not only was the earth not the centre of the solar system – our solar system was not even the centre of the galaxy. . . .

Shapley concluded – and his reasoning later turned out to be correct – that the globular clusters are associated with our galaxy. Some are inside, some are outside; and they move around the galaxy in eccentric orbits. The original 'nebulae' have proved to be what Halley supposed they were – clouds of shining gas in our galaxy or gas shells surrounding dying stars (planetary nebulae). But what about the other main types of nebulae – the elliptical and spiral ones?

The answer to this problem began to emerge in 1919, when the American astrophysicist Heber Doust Curtis observed that novae – new stars like the one Tycho had observed – were about ten times fainter in spiral nebulae than novae in our own galaxy. This made the 'average' spiral nebula about a million light-years away, and that was pretty certainly far beyond our own galaxy.

This was also a conclusion reached by Hubble, who had been studying nebulae through the 100-inch telescope at the Yerkes Observatory since the end of the First World War. Hubble's giant telescope, the largest of its time, was able to make out Cepheid variables in the Andromeda nebula, and he calculated that it must be about three-quarters of a million light-years away.

Even so, neither he nor Shapley could accept that the Andromeda nebula was as large as our galaxy – or, indeed, that any of the others were. If Andromeda (known to astronomers as M31 – No. 31 in Messier's catalogue) was three-quarters of a million light-years away, then its size had to be far smaller than ours. It seems to have struck no one as highly unlikely that our galaxy should be unusually large.

By this time, it was generally realized that the extra-galactic nebulae – which could now be referred to as galaxies – were moving away from us; spectrographic analysis revealed the typical Doppler red-shift. (Oddly enough, Andromeda is one of a few exceptions; Slipher discovered as long ago as 1913 that it is moving towards us.) This did not worry anyone unduly; to begin with, de Sitter had published his paper in 1917 asserting that, according to the formulae of General Relativity, the galaxies would be moving away from us, or would deceive us into thinking so. So it could simply be some odd Einsteinian effect. And if the Andromeda galaxy was moving closer to us, then probably others were as well.

By the mid 1920s, enough galaxies had been carefully observed to make it clear that the Andromeda galaxy was an exception; all the other galaxies – or most of them – were moving away. In 1929, Hubble published his startling observation that the galaxies are receding from us at a speed proportional to their distance. This again is a natural consequence of the Einsteinian view of space-time. If we think of the universe as the skin of a balloon, and imagine it expanding at a uniform rate, then a galaxy which is, say, an inch away from us on the skin will double its distance eventually to two inches. But this means that a galaxy that was previously two inches away will double its distance to four inches in the same time, and one that was eight inches away will double its distance to sixteen inches. So although the skin itself is expanding at the same rate, more distant objects will be receding at a greater speed than those close to us.

Eddington's ex-student, the Abbé Georges Lemaître, was the first to follow these observations to their logical conclusion. If the universe is now expanding – or exploding – there must have been a time when it was, so to speak, an unexploded bomb – a cosmic egg, or primeval atom, or 'singularity', or whatever term we happen to prefer.

When did this 'big bang' occur? From Hubble's observations about the recession of the galaxies, it looked as if it must have started about two billion years ago. But that seemed an absurdly short time. Study of radium and other radioactive elements in the earth's crust indicated that our own earth was older than two billion years.

The puzzle continued to defy solution until 1942. The man who stumbled upon the solution was a German-born astronomer, Walter Baade, working at the Mount Wilson Observatory, twenty miles north of Los Angeles. When the Mount Wilson Observatory was built in the early twentieth century, no one realized that a small town called Hollywood would expand to join Los Angeles, and that the resultant smog and the glare of its lights would make astronomical observation difficult. But in 1942, the War plunged Los Angeles into darkness. And Walter Baade was able to use its magnificent telescope to study the Andromeda nebula. He was able to observe that the stars in the centre of the nebula were quite different from those in its spiral arms. The outer stars – which he called Population I – were bluish in colour, and many of them were far brighter than our sun; those of the inner galaxy – Population II – were reddish in colour. The blue stars were younger and more vigorous; the red ones were much older. Further studies with a 200-inch telescope at Mount Palomar revealed that most of the stars in the universe are of the older type. Moreover, most of the galaxies in the universe are not spirals, but ellipticals, and are made up mostly of these old stars. Spiral galaxies (like Andromeda), on the other hand, have red stars in the middle and blue stars in the arms, which are also full of cosmic dust. These blue stars – Population I – are only a few per cent of all the stars in the universe. Our sun is one of them, and so are many of the stars in our immediate neighbourhood. It followed that our galaxy is a spiral, and that our sun is in one of its arms. One of the problems of astronomers is the number of dust clouds in our neighbourhood. . . .

But this also meant there were two types of Cepheid variables – red and blue. The Cepheids that Shapley and Hertzsprung had measured in our galaxy were reds, and they had correctly determined their distance (more or less). But Shapley and Hubble had based their assessment of the distance of the Andromeda nebula on the assumption that the Cepheids they could make out in its spiral arms were the same as the ones they had observed in our galaxy. This was untrue. They were many times brighter – up to a hundred times. So an estimate of distance based on these would be far too short. When corrected, it was clear that the Andromeda nebula is about two million light-years away, not three-quarters of a million; and its size is slightly larger than that of our galaxy, not smaller. It followed that the other galaxies – which Hubble had judged much smaller than ours – were also at least twice as big as Hubble had thought.

In one stroke, Baade had doubled the size and age of our universe.

CHAPTER NINE

IN THE BEGINNING

From the point of view of the astronomer, it is a pity that our earth is situated in such a remote backwater of the Milky Way galaxy. We are on the middle arm of three great spirals – known as the Orion arm – and the consequence is that when we try to study the centre of our galaxy, we see only a giant cloud of dust, like a fog-bound city. It is so dense that even the most powerful telescopes cannot tell us what is going on. And if the telescope was the astronomer's only way of penetrating interstellar space, we would have to reconcile ourselves to permanent ignorance.

In the early 1930s, Bell Telephone Laboratories were steadily improving short-wave radio-telephone communication between remote parts of the world; but they seemed unable to solve the problem of static interference. A young engineer named Karl Guthe Jansky was asked to investigate, and he rigged up an aerial to scan the skies at Holmdel, New Jersey. Eventually he pinned down the source of most of the major causes – thunderstorms, electrical apparatus, aeroplanes – but one steady hissing noise continued to elude him. It seemed to be coming from the sun, and at first he assumed it *was* the sun – whose sunspots produce radio waves. But sunspots were quiet that year. Besides, the source of the waves gained on the sun by four minutes every day.

It was this figure that gave Jansky his clue. For the starry zodiac gains on the sun by four minutes a day – which is why the sun moves from constellation to constellation. By the following spring, Jansky had calculated that the waves were coming from the constellation Sagittarius – which is in the direction of the centre of our galaxy. The radio waves, it seemed, were penetrating the cosmic dust clouds as easily as they could penetrate fog.

But astronomers in the 1930s were more interested in galaxies than in weak radio signals of unknown origin. They could not see

The galaxy M82 in Ursa Major has a disturbance at its centre. Is this a 'big bang'?

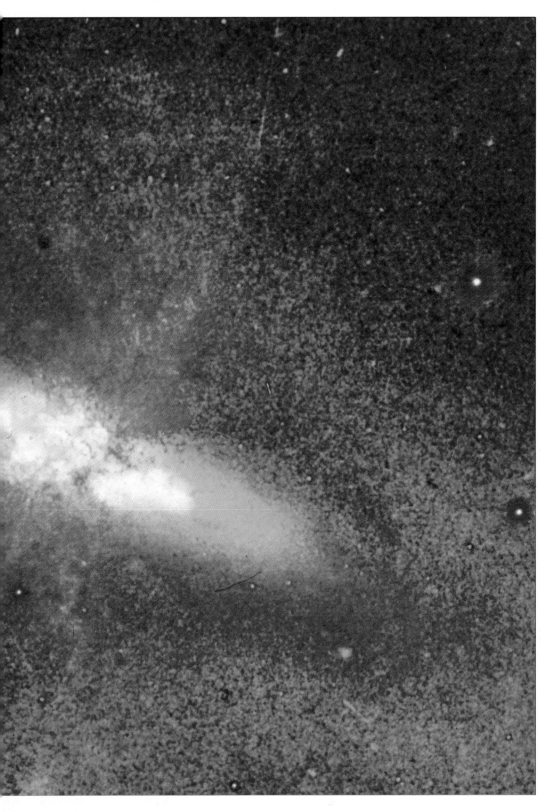

where such research might lead. So Jansky's discovery was virtually ignored. One of the few to follow it up was an amateur named Grote Reber, who built a 'dish' aerial in his own backyard in Illinois, and made maps of the Milky Way which showed radio emission from various places.

During the War, a British engineer, James Stanley Hey, studied the methods the enemy were using to jam radar, and discovered that the sun was often responsible. Like Reber, Hey began mapping the Milky Way, picking up its unexpected radio signals. It was his work that stimulated the British astronomer, Bernard (later Sir Bernard) Lovell, to build the world's first giant radio telescope at Jodrell Bank.

But what was happening to produce these strong radio waves? Where our own solar system was concerned, the question was easy to answer. Its most powerful radio emitters are Saturn, Jupiter and Venus – all planets with a turbulent atmosphere of seething gas. And a sunspot is also seething gas with a powerful magnetic field. But compared to the rest of the universe, even the sun is a feeble transmitter. In our own galaxy, the two most powerful radio sources are the Crab nebula, and Cassiopeia A.

But these are caused by supernova explosions. When a massive star has used up almost all its energy, it becomes a 'red giant', until there is finally not enough energy flowing out to resist its own gravitational pull. Its core dramatically collapses, and unsupported, the overlying layers fall in. The rapid heating results in a tremendous explosion, that sends its outer layers flying off into space at a rate of millions of miles a day. The Crab was a star which exploded in 1054 – or rather, it was seen to explode by Chinese astronomers then; the actual explosion had taken place six thousand years earlier. And even now, nearly a thousand years later, the Crab nebula *looks* like a tremendous explosion. Its giant cloud of gas – still expanding at eight million miles a day – sends out the radio waves.

Yet even these waves are not powerful enough to reach far beyond our own galaxy. It was in 1951 that astronomers began to realize that far more violent events take place in other galaxies. Reber had reported a powerful source of radio waves in the constellation Cygnus. Baade turned the 200-inch telescope on it, and found an odd-shaped galaxy with a double centre, a little like two fried eggs that have mixed together in the pan. Could these be two galaxies in collision? His spectroscope – which revealed large quantities of hot, excited gas – certainly seemed to support this idea. Of course, colliding galaxies do not annihilate one another; their stars are so far apart that they can pass through one another with relatively few collisions. But their cosmic gas clouds collide at a rate of thousands of miles an hour, which results in rapid heating.

For a time, astronomers were inclined to believe that these immensely powerful extra-galactic radio sources were probably caused by collisions between galaxies. (In certain parts of the sky, galaxies are packed fairly closely together, sometimes with only their own width between them.) But as radio telescopes located more and more of these galaxies, this explanation began to seem far-fetched – galaxies are bound to collide occasionally, but not with the frequency of dodgem cars in a fairground. It seemed more likely that galaxies were actually exploding. But why should they? Fred Hoyle put forward a possible explanation. In the centre of a galaxy, stars are crowded together; so an exploding supernova could trigger off another, and then another. It is even conceivable that one supernova could trigger off two, and these could trigger off four, and so on, like an atomic chain reaction. But if this happened, the sheer mass of the exploding stars would probably cause them to collapse into a 'black hole' – a concept we shall examine in a moment.

The radio sources we have been discussing so far are immense clouds of stars and gas. But more sensitive radio telescopes revealed that there seemed to be another type of radio source, far smaller – more, in fact, like an individual star. The early radio telescopes lacked the precision to focus on anything as small as a star. Improvements in their design began to reveal more and more of these 'radio stars'. Many of them had

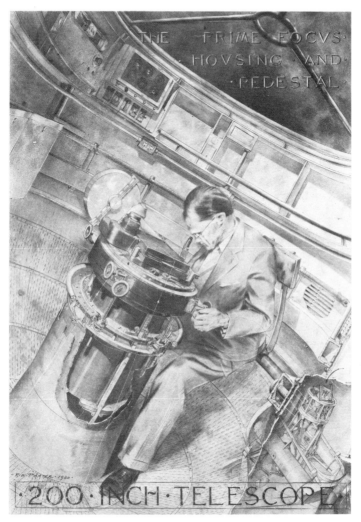

The final adjustments. Photography by means of a 200-inch telescope

been known for a long time, but had been assumed to be ordinary stars inside our own galaxy. Now, because of their powerful radio emissions, they were studied more closely by optical astronomers, and their spectra were examined. The result was baffling. These spectra resembled no known element. Then, in 1963, the Dutch-American astronomer Maarten Schmidt had an inspiration. The pattern of lines in the spectra looked like hydrogen, but the emission lines were too far down the spectrum. Could this be a typical Doppler red-shift? If so, it was enormous – so enormous that, according to Hubble's theory (that the furthest objects in the universe are retreating the fastest), they had to be further away than any known galaxies, and receding at an almost unbelievable speed – in some cases, 90 per cent of the speed of light. Could they be galaxies? Closer examination ruled that out – they were too small. But how small were these quasi-stellar objects – or quasars? Since they showed variations in brightness in the space of a few years, they had to be no larger than a light-year or so in diameter; some calculations suggest they be as little as a 'light-week', 100000 million miles. But if a quasar is a single object, even this size is staggering. The largest star we know (VV Cephei) is only a thousand million miles in diameter – a hundred times smaller than a quasar.

If their red-shift was to be trusted, these were the most distant objects known. Two

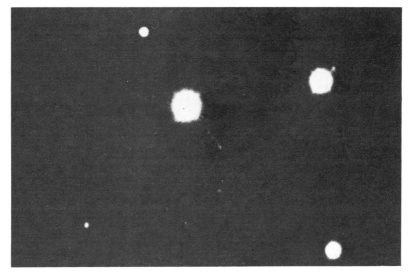

Quasars are the active nuclei of distant galaxies. This is 3C273 in the constellation Virgo
Overleaf: Computer enhanced picture of the M87 galaxy which has a distinctive jet
and is a prime candidate for a black hole

of them have been estimated at a distance of over ten thousand million light-years.

The current view is that quasars are bodies that existed in the early days of the universe, when matter was much more compact. But many mysteries remain. They are emitting energy at such a fierce rate that they cannot possibly last more than a million years or so. And since we are now observing them by light that left them billions of years ago, what have they become since then? Here a clue may have been provided by another student of quasars, Allen Sandage, who discovered in 1965 some distant blue 'stars' with no radio emission; there are even more of these 'blue stellar objects' (abbreviated to BSOs) than quasars, so that they *could* be burnt-out quasars.

As for the other question – what quasars were before they became 'cosmic gushers' – the answer could be provided by the Finnish astronomer T. Jaakkola, who has suggested that a quasar could be a galaxy *in the process* of being born. And this idea is less far-fetched than it seems.

We still know almost nothing about the true nature of quasars; but the discovery of these 'cosmic gushers' made one thing clear: that our universe is a far more *violent* place than earlier astronomers had assumed. We have become accustomed to the idea of our green earth spinning peacefully through space, entering an ice age every ten thousand years or so, but otherwise staying much the same over the millennia. We are, in fact, in the midst of an explosion, and it just happens to be an amazing piece of luck that we have found a quiet corner of it.

Another discovery of the 1960s underlined this view of our violent universe. At Cambridge, Professor Antony Hewish had supervised the construction of a radio telescope capable of detecting very short bursts of radio energy. Almost immediately – in 1967 – his research assistant Jocelyn Bell detected a radio source that was pulsating with absolute regularity, like a lighthouse beacon. For a while, the team thought they might have picked up deliberate radio transmissions from some extra-terrestrial civilization. Fortunately, they kept this to themselves – it would inevitably have become a headline story – and decided to check whether it was unique. It was not. In fact, there proved to be dozens of others. They were, inevitably, christened 'pulsars'.

Now pulsars are nothing like quasars; these are radio sources *within* our galaxy. What makes them so odd is that they pulse so fast. To be emitting regular flashes at such a pace, the pulsar must be spinning on its axis like a top. One pulsar in the Crab nebula was 'flashing' thirty times a second. It had to be a small object, but tremendously dense, tremendously hot, and with an immense gravitational field. A dying star or 'white dwarf'? Even that would not be dense enough. It looked as if Miss Bell had found something that Robert Oppenheimer had predicted in 1939 – the 'neutron star' – a star so dense that it consists entirely of neutrons packed shoulder to shoulder.

This requires some explanation. A star about the size of our sun begins as a huge ball of stellar material, which is drawn together by gravitational forces; it 'condenses', and so becomes a star, blazing with the heat created by nuclear reactions taking place in its dense core. After ten thousand million years or so, its nuclear fuel almost exhausted, it turns into a 'red giant', with a diameter fifty times as great as our sun, and a luminosity more than a thousand times greater. It becomes unstable; very gently, it puffs off its outermost layers, and for one brief moment of cosmic time, it is a planetary nebula. As the wisps of gas disperse, the dense, collapsed core – once supported by the outflow of energy – comes into view. This minuscule remnant, only the size of our earth, is so compressed that a cubic inch of its matter would weigh a ton. As a white dwarf star, it is doomed to a lingering death, leaking away its heat into space.

But if the star is more massive than our sun – say three times more massive – the core collapse will be even more violent, as gravity is stronger and electrons and neutrons are jammed together into a solid mass. This is a 'neutron star'. It could be only fifteen miles in diameter, but a cubic inch would weigh millions of tons.

The more the star contracts, the faster it spins. And the rate of these pulses suggested that this could be nothing but a neutron star. Oppenheimer's prediction had been justified. Which reminded scientists that he had also made another speculative prediction at the same time: that if the original mass of the star was greater still – say ten times that of our sun – the gravitational pull of the final neutron star could be so great that even light could not escape from it. It would become, in effect, invisible – a 'black hole' in space. More recent research in sub-atomic physics has suggested that there are even smaller particles which make up the proton, and the physicist Murray Gell-Mann has named these hypothetical particles 'quarks'.* Is it possible that quark stars exist?

* He makes it rhyme with fork; but James Joyce, in *Finnegan's Wake* (from which Gell-Mann took the word), makes it rhyme with mark, bark, dark and park – the syllables represent the cry of gulls.

241

*Giant radio telescope in Australia. The area of the dish is
three-quarters of an acre*

Do black holes really exist? *If* the theory is correct, they should be fairly common; for at one time or another, our galaxy must have contained a thousand million stars sufficiently large to collapse beyond their Schwarzschild radius (the radius at which gravity becomes so powerful that light cannot escape); so our galaxy should be littered with them, forming a possible hazard for future interstellar travellers. But since no light can escape from them we have no direct means of detecting them. However, this would not be true of black holes associated with another star (forming a binary, like Sirius). Here there would be very definite effects. The black hole would suck matter away from its companion like some giant vacuum cleaner, and this matter would swirl into the black hole, to be 'annihilated' in its turn. But since the hole is fairly small, the matter cannot be sucked in as fast as it is pulled off the companion star; it would be pulled into the 'throat', and the resulting compression would be so powerful that tremendous energies would be produced – energies within the X-ray range.

Now X-rays cannot penetrate the earth's atmosphere, which means that they cannot be detected in our observatories. But the development of rockets and satellites has made it possible for astronomers to send their equipment out into space, and they have been doing this for the past two decades. The original intention was to study the X-rays that emanate from the sun's corona, the mantle of thin gas surrounding the sun. But the instruments soon detected far more powerful X-ray sources elsewhere.

The great gas clouds of the Crab nebula emit X-rays; so does a stellar object in the constellation Scorpio. The X-ray emissions were early evidence of the existence of neutron stars, before they were confirmed by the discovery of pulsars.

X-ray astronomy has produced several possible candidates for black holes. The X-ray source in Cygnus, called Cygnus X-1, is a supergiant star, twenty times as heavy as the sun, with a five and a half day period. This tells astronomers that it has a companion ten times as massive as our sun. But the X-ray pulses indicate that its companion can only be twenty thousand miles across; and these figures suggest a black hole.

Astronomers also speculated that black holes could form the core of some globular star clusters – that is, that a star cluster (which possesses a high density of material for its size) could begin as a great ball of gas, whose centre could bè dense enough to collapse into a black hole. The same process may occur in the formation of galaxies; and in 1978, the galaxy M87 in Virgo presented itself as a likely candidate. A group of astronomers at the Hale Observatories detected a great number of stars at the centre of M87, apparently held together by some invisible mass. Other observatories succeeded in measuring the mass of this invisible 'hole' with the help of an instrument called an Image Photon Counter (developed by Alec Boksenberg of University College, London), which amplifies the faint light signals from around the suspected black hole, and enables the speed of the stars to be measured from their spectral lines. They concluded that the mass required to hold the stars 'in place' at that speed would be five thousand million times the mass of our sun. Again, the likeliest explanation is a black hole.

It has been plausibly suggested that black holes may be the explanation of quasars. A black hole gobbling up matter from surrounding stars would only need to absorb a mass equivalent to our sun every year to blaze at the observed rate. There is a rather more disturbing observation which suggests there may be an immense black hole at the centre of our own galaxy – to account for the infra-red radiation coming from that direction.[1] The physicist Joseph Weber, from the University of Maryland, is also convinced that he detects powerful gravitational waves coming from the same direction – disturbances in the galaxy's gravitational field. But his figures suggest that if a black hole was responsible, it would have to swallow up thousands of stars every year. So far, no one has managed to duplicate Weber's results.

The real problem in writing about black holes is that we are entering an area where science begins to border on science fiction. We are told that once matter has passed the Schwarzschild radius, there is nothing to stop it from collapsing indefinitely – although no one has so far explained on what evidence science assumes matter to be infinitely compressible. What, then, would happen to matter so compressed into the infinitesimally small? It has been seriously suggested that the black hole may be an entrance into another universe, and that if we could somehow enter that other universe, we would see the reverse of a black hole – a 'white hole' or cosmic gusher – spewing out matter and energy, apparently creating it out of nothing. If black holes had been suspected in 1950, the idea would have offered support to the Steady State theory of the universe, propounded by Fred Hoyle, Herman Bondi and Tommy Gold, who argued against the Big Bang theory on the grounds that it still left unanswered the problem of what caused the bang. They made the alternative suggestion that the reason the universe is expanding is that matter is being continually *created* – out of nothing – and that this constant change in the amount of matter in the universe causes it to expand. For a while, the Steady State theory was given serious consideration. But in 1965, it received its death blow.

At Holmdel in New Jersey (where Jansky discovered radio sources), two astrophysicists, Arno Penzias and Robert Wilson, were tracking echo satellites with a vast metal horn. Their instruments picked up a weak radio signal, a background hiss, which at first they assumed to be caused by some defect in the horn – or possibly by the birds who made a habit of nesting in it. The physicist George Gamow (one of the main proponents of the Big Bang theory) had predicted that if the Big Bang occurred thirteen billion years ago, the gamma radiation produced by it ought still to be detectable – but with a Doppler-shift that would stretch it to radio waves. At the time Penzias and Wilson discovered the background radiation, Robert Dicke and a team at Princeton were building a special radio telescope to look for the waves predicted by Gamow. They were surprised and delighted to learn that the Holmdel team had already found them. Since that time, the Big Bang theory has held the field without rivals. The 'white hole' theory has come too late to save the Steady State universe.

There is, nevertheless, an interesting alternative theory that combines basic characteristics of the Big Bang and Steady State theories. It was put forward by the mathematical physicist Paul Dirac – the man who first predicted the existence of anti-matter (or at least, of the positron, the opposite of the electron). Dirac's theory is based on what he calls the 'big number hypothesis'. He is intrigued by the problem of why nature seems to favour certain ratios – for example, the ratio of the mass of an electron and a proton (about 1000), or the ratio of the charge of an electron and Planck's constant (137). And he points out that two very large ratios (he calls them 'dimensionless numbers') are practically identical. One is the age of the universe expressed in the 'atomic unit' of time – the time it takes light to cross an atomic nucleus; Dirac says this is ten to the power of thirty-nine – a one followed by thirty-nine noughts. The other is the ratio between the electric force and the gravitational force inside an atom containing one proton and one electron, which is once again 10^{39}. If, says Dirac, this is no coincidence, and the two numbers are connected by some causal law, then it would follow that gravity is not a constant. For as the universe gets older, one of these numbers will increase, and the other – involving the gravity between an electron and proton – must increase to keep up with it. Gravity itself should therefore decrease. But another vast number that must increase is the number of particles in the universe; and this demands 'continuous creation' of matter, just as in the Hoyle-Bondi Steady State theory. Nevertheless, Dirac accepts that our universe began as a 'singularity' (a single point), and that it will continue to expand forever.

It may seem that these various theories of the universe are not only bewildering but self-contradictory. But the differences are more apparent than real. At the beginning of the eighties, it can be said that scientists are in broad general agreement about the main facts of cosmology. To begin with, all present theories are based upon Einstein gravitational formulae (although Dirac confesses to certain similarities between his own ideas and those of the late Professor E. A. Milne, who rejected General Relativity as unnecessary). And then, the basic facts – the universal red-shift, the 'background hiss' – hardly admit a broad range of interpretation. Let us try to arrange the known facts – and speculations – into a general pattern.

Computer enhanced picture of the spiral galaxy NGC 1097

Between nine and eighteen billion years ago the universe consisted of one gigantic fireball, which may already have existed for an indefinite period in a state of equilibrium, like our own sun. Radiation could not escape from it, because it was, in effect, an immense black hole, with a temperature of about a trillion degrees. Its size, according to an estimate of Professor Lloyd Motz,[2] was about a billion miles. At such immense temperatures and pressures, only very basic particles – like protons, electrons and neutrinos – could exist. There would also be anti-electrons, anti-protons and anti-neutrinos; these particles can now be created in the laboratory in high energy accelerators, which cause collisions between particles. When a particle and an anti-particle meet one another, there is a flash, and they both vanish. And in the original fireball, which was a kind of super-accelerator, particles would continually be appearing and vanishing.

Then came the bang. Why? This is a point that all the theorists gloss over. Clearly, there can have been no 'outside intervention', since there was no 'outside'; and it is difficult to see why the inner state of equilibrium should have been disturbed. When challenged, astronomers reply that it is not their business to explain why the big bang happened; they can only point to the expanding universe as evidence that it did. (Dirac himself for a while rejected the idea that the universe is expanding, and argued that the red-shift is caused by the change in the force of gravity.)

When the big bang occurred, vast quantities of particles and anti-particles were liberated. In the first three seconds, major changes occurred; the temperature dropped to five billion degrees. Basic particles annihilated one another and their varying amounts altered from split second to split second. (One physicist, Steven Weinberg, has even devoted a whole book to the first three minutes after the big bang.) After a few minutes, neutrons and protons came sufficiently close together to link up in groups of four – two protons, two neutrons; the combination we know as the alpha particle. It is also the nucleus of the helium atom. Sometimes, proton and neutron captured one another, making deuterium, an isotope of hydrogen. (An isotope is an element with a few additional neutrons in its nucleus.) Other simple elements like lithium, beryllium and boron were also formed in tiny quantities. At this stage, no more complex elements could be created, because the powerful nuclear force – the strongest force in nature – operates at a tiny distance of a trillionth of a centimetre, and the chance of more than two or three protons and neutrons coming into collision at the same moment is infinitesimal – except under immense pressures.

So now we have an expanding universe consisting of about 75 per cent hydrogen and 25 per cent helium, its rate of expansion being slowed down by gravity, and the operations of the laws of chance bringing clumps of gas together into clouds, whose gravitational force attracts more gas. This part of the story sounds very much like the nebular hypothesis of Kant and Laplace (and again we have to marvel at the intuition displayed by the German philosopher more than two centuries ago). After a billion years, the temperature drops to about 5000°, while the diameter is about a thousandth of the present-day universe. Great clouds of hot gas form into galaxies – there is still a great deal of argument about why some are elliptical and some spiral – and the whirls and eddies in these spinning islands of gas contract into stars. And now, in these vast droplets – perhaps a million miles in diameter – the pressure is at last great enough to form the next element in the chain: carbon, with its nucleus of six protons and six neutrons. From the human point of view, this is the most important of all the elements, since it is the basic 'building block' of life.

But life cannot exist without various heavier elements – iron, phosphorus, sulphur – and these were also formed in the core of the original giant stars. So in this early universe, a few billion years old, there was no possibility of life; for even in these pockets where the temperature was low enough to support life, the essential elements were missing, locked deep inside the stars at millions of degrees. Eons of time had to pass, until the original stars grew old and collapsed, hurling the heavy elements out into space in the form of dust clouds. And it was the new stars that formed from these gas and dust clouds, whose planets became the cradle of life.

And how did life evolve? This is still a matter of some controversy. The obvious explanation is that the various elements – carbon, nitrogen, phosphorus, oxygen, hydrogen, iron – somehow came together in the great witches' cauldron of our cooling planet, fused together by lightning, and formed the complex molecules called amino-acids, the first stage in the history of living organisms. But in 1963, radio astronomers discovered molecules of combined oxygen and hydrogen out in deep space; these 'hydroxyl groups' only need one more atom of hydrogen to become water, the basic necessity for life. And six years later, they actually discovered water (by its spectrum), as well as ammonia, formaldehyde and methyl alcohol. The latter contains six atoms. Since then, molecules containing up to nine atoms have been discovered. The simplest amino-acid, glycine, has ten atoms. No one is quite certain yet how these molecules came to be formed; but the one thing that is certain is that they are the basic chemicals of life. In their book *Lifecloud*, Sir Fred Hoyle and Professor Chandra Wickramasinghe have gone on to argue that these 'building blocks' of life were brought to earth by comets. And then – presumably – the lightning did its work of fusing these into more complex molecules, and the warm seas did the rest.

This is, admittedly, a rather odd part of the story. We can easily see how heavy elements were built up in the interior of stars, and the presence of simple molecules in space is hardly an impenetrable mystery – presumably they were ejected from exploding stars and built up slowly on the surfaces of cool dust grains. But how did these simple 'building blocks' turn into organic matter? Hoyle admits that 'we would have a straightforward answer to this question if we could reasonably argue that the initial arrangement of a living cell was due to chance'.[3] But, he goes on, that is practically impossible, because the least complex protein chain necessary for life is made up of a hundred amino-acids linked in a chain, and with twenty possibilities for each link in the chain, the number of possibilities amounts to a staggering figure – a two followed by a hundred noughts. And even if new combinations were tried every few seconds, it would take millions of times the lifetime of our earth to try them all. In fact, even if all the hundred million trillion possible planetary systems in the known universe experimented busily at the same time, there would still only be time for a tiny fraction of the necessary 'trials' – 10^{80} as compared to 20^{100}. Which might seem to suggest that something apart from mere chance was at work – some kind of purpose. But Hoyle carefully sidesteps this conclusion, and only speaks of 'a compact association of physical laws, environment and life, an association that would bring together the disciplines of the physicist, the astronomer, the geophysicist and the biochemist'. He fails to explain how this association could complete the necessary permutations in a sufficiently short period of time.

At all events, it would seem that life appeared on our planet in an incredibly brief period of time, and proceeded to build itself up into more and more complex units. And, in the past half billion years, the phenomenon called consciousness appeared.

We are so accustomed to thinking that to be conscious *is* to be alive that it is important to emphasize that consciousness is a relatively late and relatively feeble invention of life. Trees and bacteria seem to survive very well without what we would call consciousness. Moreover, most of our own important vital processes operate without its aid. This unconscious force of life, which organized the miracle of the eye and the brain, seems to 'know' infinitely more than the most powerful human intellect. It seems to have evolved consciousness as a useful addition to its survival equipment, since a conscious creature can exercise foresight to avoid danger and create conditions of security.

Finally, a mere two or three million years ago, man appeared. At first, it was all he could do to keep up with the other animals with whom he shared the savannahs and forests. In that curious and stormy period known as the Pleistocene – which began about a million years ago – he began to show definite signs of his future dominance; he became more than a survivor, and developed into a predator. He learned to use simple weapons. He became increasingly a social animal. Yet, as we have seen, he remained basically a child of nature, a creature of the earth. When Cro-Magnon man exterminated his more 'instinctive' brother, the Neanderthal, a mere thirty-odd thousand years ago, the new lord of the earth was still aware of his place in nature – even though he now made conscious attempts to dominate it through the force of magic. But if Julian Jaynes is correct, this awareness of nature began to give way to a new kind of consciousness only about five thousand years ago. By 1000 BC, man had become 'stranded' in consciousness, isolated from the rest of nature, like a hermit living on the top of a high pinnacle. The age of science had begun.

And, of course, books like Hoyle's *Lifecloud* and *Ten Faces of the Universe* are typical – and very valuable – products of this age of science. No one is going to quarrel with what they say, which is stimulating and imaginative. The trouble lies in what they don't say. The picture that emerges from Hoyle is much like the picture that emerges from Motz's *The Universe, Its Beginning and End*, or John Gribbin's *White Holes: the Beginning and End of Space*, or John Taylor's *Black Holes: The End of the Universe?* These subtitles share the same apocalyptic note, and in view of their contents, this is understandable enough. Motz surveys man's own destruction of his environment – for example, the destruction of the ozone layer by jet aeroplanes – and points out that if our thin cover of ozone disappears, we shall all succumb to skin cancer; and this is only one of a dozen possible kinds of extinction that man could bring upon himself unless he changes his ways. Motz then points out that even if humanity can avoid these dangers, the friction of the tides will eventually make the earth rotate more and more slowly, until the moon finally disintegrates and spreads around us like the rings of Saturn or Uranus. We have no immediate cause to worry, since this will take a few billion years. If we survive the destruction of the moon, we shall have to find the means to save ourselves from being roasted alive by the sun, which will become larger and far hotter than at present. Our oceans will turn to steam; and the sun will turn into a red giant. We may, of course, succeed in colonizing another planet; but it may have to be in another solar system.

It is also true that we shall eventually find our galaxy more or less alone in the universe (apart from a few local galaxies like Andromeda, which are not receding from us); but this hardly seems to matter. If we could survive the death of our sun, the absence of other galaxies would hardly concern us. If our own galaxy can survive then we shall eventually learn the answer to that interesting question: whether the amount of matter in our universe is great enough to cause it to stop expanding, and to contract instead. But again, it is doubtful whether we shall survive that long, since by that time all the energy in the universe will have thinned out to such an extent that we shall all die of cold. Physicists express this by saying that the entropy in the universe – the random element that increases, for example, when you break something – is continually increasing, so that the universe is inevitably running down.

So from the strictly scientific point of view, long-term prospects are not merely bleak, but non-existent. Laplace told his contemporaries that the universe was stable and unchanging; even Einstein thought so when he published the General Theory of Relativity in 1915 (although Slipher had discovered the recession of the galaxies two years earlier). Within three years, de Sitter had convinced Einstein that he was wrong. And for the past half century, most scientists have accepted that we are living in a doomed universe. It may cease to expand; but if it does, it will then contract until it collapses into a huge black hole. Then, conceivably, it will explode again, and our universe will once more come into being. Even in an 'oscillating universe' like this, nothing human can possibly survive. . . .

But this vision of ultimate destruction may leave many things out of account. To begin with, there is that figure of twenty to the power of a hundred – Hoyle's estimate of the number of permutations possible in the amino-acid chain. Science can explain how the nuclei of heavy elements came together under the tremendous pressures inside stars; but it cannot provide a parallel explanation for the complexity of organic molecules. In an image I have used elsewhere, it is like asking us to believe that dozens of bits of rusty cars in a junkyard could combine 'by chance' to form a new Rolls Royce. Until a better explanation can be devised, it seems to me logical to suppose that life itself was responsible for the organization, and that it proceeded by some method other than trial and error. And here it may be useful to recall the calculating prodigies of our first chapter, who could solve 'insoluble' mathematical problems – like whether the Fermat number was a prime – by a form of direct intuition. Science feels obliged – by the rules of the game – to rely entirely on the intellect and the left brain; it sees the universe through the intellect and the left brain. Which means that it feels obliged to ignore any form of knowledge that short-circuits its deliberate procedures. And where the organization of life is concerned, it finds itself faced with a problem that defies solution in terms of these procedures.

It seems, then, permissible to theorize that the force of life played an active part in organizing the endless complexities of the amino-acid chains and the DNA chains, and that it did so by some method other than trial and error. 'Unconscious life' seems to have a very clear idea of how to organize itself. A typical example is cited by the biologist Sir Alister Hardy, who describes how the *Microstomum* worm eats a polyp called the *Hydra* for the sake of its stinging capsules. When the *Hydra* has been digested, the stinging capsules are filtered through the stomach of the worm, and carried by special cells to its skin, where they are mounted in the 'ready' position to discourage predators. All these operations are carried on 'unconsciously', automatically; but it is difficult to see how remote generations of flatworms learned the trick by trial and error, in the way that a man fumbles his way around a strange room in the dark. It looks rather as if the 'unconscious' processes know where to find the light switch.

But if we allow this assumption of purpose – a word that makes all good scientists shudder – then it is hard to see why we should abandon it when we come to describe more complex forms of life, like birds and animals. The Darwinian view tells us that life played no active part in its own evolution; it merely waited passively for chance to provide it with opportunities. When nature was generous, everything flourished; when times became hard, only the well-adapted survived. The chance of survival was as remote as winning the football pools; but there were enough winners to allow life to continue and evolve. The bear that happened to be born white – by some freak of the genes – survived the snow, and reproduced its own kind.

But if life organized its own foothold on this planet, it seems more logical to assume that it may have played a more active part in its own evolution. That is, that the unconscious will that carries the stinging capsules from the stomach of the *Microstomum* may have played some part in whitening the fur of the polar bear and lengthening the neck of the giraffe. It may play only a small part in the evolution of any thousand generations – so little that biologists are convinced of its non-existence; yet it seems that, where matters of life and death are concerned, it may intervene decisively.

If we are willing to consider this view of life as a highly adaptive and highly intelligent force, then the conclusion of the current generation of cosmologists no longer seems inescapable. Instead of seeing the running down of the universe – and the disappearance of life – as a purely automatic process, it is conceivable that life will continue to take care of itself as resourcefully as it has in the past half billion years. And, moreover, that if it can continue to evolve at the same rate as in the past half billion years, then such problems as the slowing down of the earth and the disintegration of the moon should present no real danger. Life is basically the harnessing of energy; and in this sense alone, man has evolved more in the past century than the horse or dog in the past fifty million years. It may be as impossible for us to imagine what our descendants can achieve as for Shakespeare to have imagined space travel and atomic power.

We must also consider the possibility that our current notions about quasars, black holes and the expanding universe are quite simply wrong. Dirac expressed his doubts about the expansion of the universe. Hoyle has expressed his doubts about black holes, as well as about the Big Bang. Speaking of quasars and radio galaxies, he has said: 'The behaviour of [these] sources cannot in any case be explained in terms of conventional physical theory. According to conventional theory, large masses in small volumes plunge into singularities [black holes]. . . . I would expect the sources to fade away . . . in a time scale of the order of a year. Instead we observe violent outbursts in a time scale of the order of a year. The observed properties are the exact opposite to what we would expect according to conventional physical theory.'[4] Later data has led to a refinement of our ideas; nevertheless the position of astronomers today brings to mind the position of Galileo's contemporaries after the invention of the telescope. New worlds were revealed; the Copernican theory was proved beyond doubt; yet they still believed that the correct number of the planets was seven – because God created the world in seven days – and that the stars, and nebulae, were spread uniformly through all space. Their new knowledge was like a young shoot on an old medieval oak tree; and they had no means of imagining, for example, the incredible consequences that would spring from Ole Roemer's discovery that light takes time to travel. Modern astronomy is based on Einstein's theory of gravitation; but if Hoyle proves to be correct about quasars and black holes, then Einstein's theory may have to be revised, or – it is conceivable – scrapped. The mere fact that Hoyle can question anything so apparently 'fundamental' suggests that our basic 'certainties' may be less solid than they appear. In the decade following the announcement of General Relativity, Professors A. N. Whitehead and E. A. Milne both suggested fundamental revisions of the theory of relativity. Whitehead was disturbed by the very foundation of Special Relativity – the idea that simultaneousness has no meaning unless you say *for whom*. Whitehead agreed that two space ships passing in empty space would find it impossible to decide which of them was moving, and therefore, whether two events were simultaneous or not. But he objected that in spite of this, we feel intuitively that it means something definite to say that two events are *really* simultaneous. Even where two space ships are concerned, you could find out whether one of them was travelling at a thousand miles a minute while the other was standing still – their captains could tell you.

Similarly, Dr Jacob Bronowski pointed out that if you were travelling away from a clock at the speed of light, the hands of the clock would remain 'fixed' as far as you were concerned, for the light from them would only just manage to keep abreast of you. But common sense would reply: 'Surely that does not mean that time has stood still? The clock *really* ticks on, and the watch in your pocket will show you the correct time.' Human beings experience a stubborn refusal to accept that the word 'really' has no meaning.

Milne objected to certain aspects of General as well as Special Relativity. Surely, he maintained, there is something wrong with calling the speed of light – which is just an empirical fact – a *law* of nature?[5] He added that he had never been convinced of Einstein's law of gravitation. When we say space is curved, 'what was it that was curved? . . . Still more, what was it that was "expanding" when it was stated that space is expanding? Could the space of this room be said to be expanding?' In which case, why isn't our solar system expanding?[6]

Anyone who thinks about relativity will find many similar objections. For example, according to Special Relativity, time goes slower for a man on a fast train; but for *him*, the time of the man on the station platform goes slower; it is reciprocal. So why, in the 'twin experiment', does one of them age faster than the other? Why is this not also reciprocal?[7] All and each of these objections may have their satisfactory answer; but they illustrate that relativity is not really the unquestionable dogma that most non-scientists assume. In a hundred years' time, we may wonder how any sane person could ever have swallowed the idea that 'space' is curved – when it would be simpler to say that particles of light are attracted by gravity like any other particles – or that there is no absolute time. Or we may be able to recognize that the 'Einstein revolution' was the beginning of a new epoch in human discovery, whose consequences could not be understood until the next steps had been taken. From the present perspective, the odds seem to be about fifty-fifty.

Medieval view of man's place in the Creation. Miniature from St Hildegard of Bingen's Liber Divinorum Operum

POSTSCRIPT

Bertrand Russell once said that the aim of philosophy is to understand the universe. We have followed man's attempts to pursue this objective from the incision of the Dordogne reindeer antler and the construction of the Great Pyramid to the enunciation of General Relativity and Quantum Theory. How much closer are we to achieving that aim?

The answer is plainly discouraging. We know – or think we know – that the universe started with a Big Bang. But that brings us no closer to *understanding* the universe. What we really want is some clue to the questions: where does space end? when did time begin? And the Big Bang theory leaves these unanswered. In this sense, it is no more satisfactory than the ancient Hindu belief that the world is supported on the back of an elephant which is supported on the back of an ox which is supported on the back of a boar. . . .

What we must recognize, quite clearly, is that science *as such* cannot answer these questions. The reason should be obvious. Science is concerned with that world 'out there', at the other end of a telescope or microscope. Science acts, quite properly, on the assumption that we can only increase our knowledge by studying the material universe, by making our microscopes and telescopes increasingly powerful. But seeing to the limits of the material universe leaves the mystery of infinity untouched, just as knowing what happened in the first milliseconds of the Big Bang tells us nothing of *why* the universe came into being.

Our minds try to grip the problem and fail; they slip off it or through it – it is hard to say which. The result is a deep sense of frustration, a contempt for the limitations of the rational intellect such as Goethe expressed when he made Faust say in the opening scenes, *'wir nichts wissen können'*. 'We can know nothing.'

The philosopher Fichte expressed it less negatively in the opening paragraph of *Bestimmung des Menschen (The Vocation of Man)*, where he explains that he has followed all the proper scientific procedures:

I believe I am now acquainted with a considerable part of the world around me, and I have expended enough care and effort in acquiring this knowledge. I have put faith only in the testimony of my senses . . . I have touched . . . I have analysed . . . So I am now quite certain of the accuracy of this part of my knowledge . . . I would stake my life on it.

But – what am I myself, and what is my purpose? [1]

That cry makes us aware that there is another *kind* of question, and that science seems incapable of coming to grips with it. For science is obliged to treat man as a part of the material universe. But when we experience a sense of meaning and purpose, it comes from *inside* us. If I am tired and bored, and I wonder what I can do about it, I look *inside* myself. Which is why no amount of looking through a microscope can answer the question, 'What is my purpose?'

But then, common sense tells us that science is not the only way of probing a question. Saint Augustine says: 'What is time? When I do not ask the question, I know the answer.'[2] Thought cannot grasp it; but intuition can. The same applies to the concept of 'life'. Thought can only grasp the notion of living *creatures*. To focus on 'life', we have to retreat into ourselves, towards the secret life inside us, that strange source of vitality and purpose that seems to lie in our depths.

To grasp this is to realize that *all* understanding depends on this 'descent into ourselves'. My senses make me aware of the physical world. If I want to listen to music, I must try to cut off most of my sense impressions. If I try to remember my childhood, my impressions are dim and vague; but if I can sink into a state of inner calm, I can begin to conjure up actual smells and feelings. When I am distracted by too many sense impressions, I cannot even tell what time it is; I look at the clock and the time fails to 'sink in'. Understanding is a process in which knowledge sinks *into* us.

It should also be clear that this power to descend 'into ourselves' is one of the central aims of evolution. (T. E. Lawrence said: 'Happiness is absorption.') An idiot is a person who lives entirely 'on the surface'; his consciousness is little more than a reflection of the external world. Yet even the most 'superficial' people have a basic desire to be *interested*, because interest focuses our powers and permits us to descend to deeper levels of control. This is what Shaw's Captain Shotover meant when he said his aim was 'the seventh degree of concentration'. Human evolution could be seen as the gradual increase of our power *not* to be mere reflections of the external world, our power to *decide* what consciousness deserves to be focused upon. At present, we have little choice but to waste it on whatever boring object is in front of our noses.

I am suggesting, then, that science is unable to provide answers to the ultimate questions about space and time because it can only provide the raw material of insight. This material must be drawn inside us before it can turn into knowledge. And what guarantee is there that the process of 'descending into ourselves' will one day reveal the answer? None whatever. It is conceivable that the universe itself is not ultimately 'rational', so it could not be understood by the deepest intuition. On the other hand, most of us are familiar with the sudden flash of insight, the recognition that makes us want to shout, 'Of course!', and there *are* moments when we seem to catch a glimpse of some completely different way of apprehending space and time. And – for what it is worth – the evidence of mystics suggests that it is possible to obtain insight into a completely different order of reality.

To which our common sense replies, 'Yes, but surely mystical insight has to obey the laws of logic – it cannot actually *contradict* it?' But this is another question upon which we had better keep an open mind. Einstein tells us that we must conceive of space as the skin of a balloon with no 'inside'. Professor Lloyd Motz tells us that we must not imagine the original 'fireball' in space, because space did not exist at this stage. Professor John Taylor tells us that black holes could be the entrances to another universe existing in another dimension. Dr John Gribbin advises us to think of a whole array of parallel universes 'embedded in superspace'. The Einstein equations also predict the existence of a particle called the tachyon which *always* moves faster than light, which loses energy as it accelerates, and which can travel backwards in time. (Of course, it may not exist; but his equations also predicted the existence of the positron, which, it transpired, *did* exist.) In each case, logic tells us that what is being proposed is absurd, like the grin of the Cheshire cat remaining after the rest of it has vanished, and that science ought to be ashamed of itself for telling us such dreadful lies. Perhaps it ought; but it has the convincing excuse that all these things are *mathematically* defensible. So it seems that our human reason is in some odd way inferior to mathematical logic.

It is true that there is no physical proof for any of the above statements; they may all turn out to be the wildest nonsense. But the tachyon's alleged ability to defy the laws of time may remind us that there *is* some impressive evidence for the paranormal ability called precognition, which is equally 'impossible'. And here we are dealing with matters that have been tested and re-tested in the laboratory.

In 1939, Dr S. G. Soal was investigating the possibility of telepathy by asking his subjects to guess the identity of various cards. His results were disappointing. A colleague, Whately Carington, had been carrying out a slightly different test in which subjects were asked to draw a 'target' picture they had not seen, and he noticed a strange effect. Some of his subjects were drawing the previous picture or the next one. Carington suggested that Soal should check his own results. Most of them remained negative. But one London housewife, Gloria Stewart, had made a high number of 'forward' guesses – the odds against chance were 63 000 to 1. Later, Soal discovered a young photographer, named Basil Shackleton, with an equally remarkable ability for guessing the next card, and the series of experiments he conducted with Shackleton over several years – leaving no possible doubt of his precognitive abilities – have become a classic of psychical research.

In 1962, a team at the Newark College of Engineering in New Jersey devised an even more rigorous test. The subjects were asked to guess a hundred-digit number that did not yet exist – it would be generated at random on a computer. The difficulty of the test made positive results unlikely, yet even here there was a remarkable success rate, one subject scoring as many as twenty-three when the chance expectation was three.

In fact, precognition had been repeatedly 'proved' from the early 1920s onwards through a test devised by Dr Eugene Osty, in which a professional psychic was asked to guess the personal details of a man or woman who would sit in a certain chair at some future date. A full record exists of such a test carried out by the Israeli Para-psychological Society under Dr H. C. Berendt. A Mannheim psychic named Orlop was asked to tape-record his 'prediction' and mail it to Jerusalem. Two weeks later, an audience was asked to take chairs at random in a lecture hall. The tape recorder was then switched on; Orlop's voice said, 'Seat number fourteen; a lady; height, one metre seventy to one metre seventy-five; age, forty to fifty. Within the last year, accident in own house injuring the knee by slipping on the steps. . . .' The occupant of seat number fourteen was a woman of forty-two; height, one metre seventy-two; she *had* slipped in her own home some months before, but had injured her ankle, not her knee. Perhaps the most remarkable 'hit' was the description of her profession, 'helping other people to spend their leisure time'. The lady was an actress.

Both logic and common sense tell us that time is a one-way street, and that the future has not yet taken place. If rigorous scientific tests prove otherwise, then we must conclude that our logic itself is untrustworthy, like the logic of Abbott's 'flat-landers' who could not conceive the idea of height.

To dismiss logic as completely untrustworthy would be absurd; for the problem obviously lies in our own limitations. What we really want to know is *how* it can fail to take account of certain aspects of reality.

The general outline of an answer was suggested by Immanuel Kant, who pointed out that one of the basic purposes of our minds is to impose order on the chaos of experience. Thus a man walking down a busy street would have a nervous breakdown if everything going on around him managed to get through to his 'attention'. He ignores 99 per cent of the racket, and concentrates on his immediate purposes. And evolution has developed other ways of bringing order out of chaos. For example, our eyes are extraordinary instruments that can distinguish between energy of .000385 mm and .000765 mm; they do this by seeing one wave length as violet and the other as red. If it became a matter of life and death for us to be able to see ultra-violet radiation, our eyes would extend their range and *invent* a new colour that does not yet exist. All our senses have their own ways of imposing their own kind of order. Kant even made the bold suggestion that space and time do not really exist 'out there' – that our minds have invented them, like red and violet. And if this is correct, it would certainly explain why we cannot grasp where space ends or time began – for the answer would be *inside our minds*. The only way to understand them would be to learn to descend deeper into our minds.

We are now returning to the territory covered in the opening chapter – particularly the intriguing matter of the two hemispheres of the brain. The person called 'I' lives in the left side of the brain. One split-brain patient was shown an 'improper' picture with the right half, and responded by blushing. Asked why he was blushing, he replied, 'I don't know.' His 'I' *didn't* know. If Zerah Colburn had been a split-brain patient, he would also have replied, 'I don't know,' when asked whether 4 294 967 297 was a prime number, instead of replying, 'No, it can be divided by 641.' If Orlop had been a split-brain patient, he would not have known who would occupy chair fourteen in two weeks' time; his right brain might have known, but would have had no way of communicating it.

The conclusion should be clear. If the right brain has an understanding of time – and perhaps of space – that the left brain lacks, then scientists ought to be trying to devise ways of persuading the right to enter more closely into their investigations, and of trying to understand its strange supra-logical processes. This is not an abandonment of scientific principles, but a simple recognition that they are incomplete.

Ancient man would not have needed to be told this; he already knew it. His knowledge seems to have been mainly pure intuition, and his intuition told him that man, the earth, the planets and the stars are all interrelated. But as his knowledge hardly extended *beyond* intuition, there would be no advantage to modern man in returning to this more primitive type of consciousness. Man has evolved the instrument of reflective self-consciousness, which has enabled him to penetrate remote corners of the universe. If he has become disillusioned with his instrument, it is because he has allowed it to lead him into certain false assumptions, the chief of which is that he *is* his consciousness. And since the conscious self is an inadequate neurotic who wastes half his strength in over-reacting to trivialities, this is serious. The scientist believes it makes no difference *who* looks down the microscope: what matters is what lies at the other end. This is untrue; man has to pull knowledge inside himself, as an octopus pulls its prey into its lair. The question of 'who' looks down the microscope is crucial.

So what we are discussing is not merely a question of understanding. The 'other self' appears to possess unsuspected *powers*. That is to say, *we* possess unsuspected powers – since that 'other self' is a part of us. This is why the realm of the paranormal cannot be ignored by science; for paranormal research is also concerned with unrecognized powers. Some of these, like telepathy, make no great strain on our credulity; others, like poltergeist effects or out-of-the-body experiences, are altogether more baffling. But if our reasoning is correct, they are as relevant to our understanding of the universe as quasars and radio galaxies. In fact, it seems almost certain that we cannot understand one without the other. A true grasp of the mystery of the universe will involve a true grasp of the mystery of ourselves. And until we understand these inner mysteries, the universe 'out there' will remain inscrutable and paradoxical.

It may sound as if I am ending this book on astronomy with a plea for paranormal research. This is not so at all; in a sense, the paranormal is irrelevant. I am suggesting that when I look out at the universe and ask, 'Where does it end? What does it all mean?', the question is less straightforward than it looks. For involved in that question is a presupposition about *who I am*. We answer that question by saying, 'I was born in 19—, and my parents were so and so.' As we say this, we know it is not the true answer; it is a superficial assessment. I would be getting 'deeper' if I replied, 'I am a human being, the result of about three million years of evolution.' But is that *all* I am? If so, then my questions about the universe are irrelevant; the only problems that should concern me are those bearing on how to survive and flourish.

Why should my mind have this desire to understand *beyond* the needs of survival? Why do human beings seem to have an appetite for knowledge for its own sake? If I interrogate my intuition, it tells me that the 'evolutionary' answer is still incomplete; that in a sense – the statement 'I am a human being' – meaning purely and simply a member of the human community – is untrue. Such a statement somehow 'reduces' me.

Yet if I turn to Hoyle's remarks about the preposterously short time it took for life to develop on earth, I begin to see the glimmerings of another answer. It is clearly absurd to say that if you go on adding atoms together until they have fused into a complex molecule, that molecule will become capable of self-reproduction. It is like saying that a skyscraper is more capable of reproduction than a bungalow. And suppose life did come into being through some accidental interaction of molecules, sun and cosmic rays; why should it not be content to rest passively? Why should it have been possessed of a desire to persist and evolve?

Since my intuition can make no sense of this notion, it turns to what seems to be the

259

sensible alternative: that life itself organized its own emergence – that it organized the staggering complexity of the amino-acid chains by some method other than blind trial and error, then went on to combine hundreds of thousands of amino-acid molecules into one protein molecule, as well as into the peptides needed to bind it together. Having organized itself into many varieties of simple living creatures, it found itself faced with another problem: that they were inclined to eat one another. So each individual had to turn its attention to the question of self-defence and disguise; survival of the fittest entered the picture.

Here we reach a crucial point in the argument. According to modern biology, the law of survival is sufficient to explain the whole story of evolution. But is this true? In the mid 1960s, when writing a book called *Beyond the Outsider*, I sought the help of the eminent Darwinian Sir Julian Huxley on the chapter on biology. I was particularly intrigued by the behaviour of an insect called the flattid bug, described by Robert Ardrey in *African Genesis*. The bug has evolved an interesting method of self-protection: it disguises itself as a coral-coloured flower rather like a hyacinth. At the tip of the flower there is a green bud; behind that, a row of partially matured blossoms; the rest of the flower consists of mature red 'petals'. Ardrey's companion waved his stick at the flower; it dissolved into a cloud of moth-like insects. Then they returned to their dead twig, crawled over one another's backs for a few seconds, and once again Ardrey was looking at a perfect flower – a flower that does not exist in nature.

But how could a whole colony of bugs have 'accidentally' learned to imitate a flower? According to Darwin, a creature that accidentally resembles something else is less likely to get eaten; so it reproduces its own kind, and the disguise is eventually incorporated into the species. Where an individual is concerned, this is easy to understand. But how can it work for a whole society? Were some of them accidentally born green, some half green and half coral, and the rest coral? Did they then accidentally happen to land on a dead twig in the shape of a flower, while the colony next door took no such precautions and perished? This would still not explain why they went on doing it – that is, at what point accident became settled purpose.

I put this question – together with several others – to Huxley, but his answer left me perplexed. He replied that there could be no question of 'purpose', but only of a mechanism called 'cybernetic feedback'. If I wanted to understand how this operated, I would have to read some technical books on genetics. I looked into Waddington's *Strategy of the Genes*, but found it too difficult; besides, it seemed to contain nothing that could be applied to a whole colony. I then looked up 'cybernetic feedback', but found that equally unhelpful. Cybernetics is the science of self-regulating systems – like the programme of a washing machine – and cybernetic feedback means simply that it can correct its own accidental malfunctions. But it still applies to individuals; there is no way in which a colony of washing machines could learn to disguise themselves as a railway engine. The whole notion of a *cooperative disguise* involves purpose; yet this clearly could not be conscious or intelligent purpose, since the bugs have not evolved that much consciousness or intelligence. We are left, then, with the notion of an 'unconscious' yet extremely precise purpose, operating on a level above the intelligence of any individual bug: the same purpose I feel inclined to assume in the case of amino-acids, proteins and DNA molecules.

This was a conclusion also reached by the cybernetician David Foster[3] in considering the idea of evolution as a process of cybernetic feedback. He pointed out that although a cybernetic system – whether a computer or an oak tree – can be 'self-governing', its original programme must be imposed by an intelligence greater than that of the system itself. The washing machine or computer cannot out-think the person who is programming it. So, if a gene is regarded as a more complex form of one of those plastic wafers that programme a washing machine, we are still faced with the problem of who devised its programme.

There is an additional complication. The more delicate and complex a process, the higher the energies needed to programme it. You cannot drill teeth with a pneumatic drill or repair a watch with a hammer and chisel. Foster states his own belief that no known energies on earth are sufficiently delicate or powerful to 'programme' the gene, and that in order to find such energies, we may have to look to outer space, to cosmic rays. In Foster's universe, the intelligence appears to be conscious.

This assumption seems to me unnecessary. The notion of an unconscious but highly purposive force is enough to provide a 'minimum working hypothesis'. This force seems to be able to operate in space and time, but is not, apparently, governed by them. And why should it wish to create ego-consciousness, whose isolation from instinct renders it clumsy and unsure of itself? The answer that suggests itself is: because instinct is *too* sure of itself, and therefore lazy and repetitive. Man has achieved more in three thousand years of bicameral consciousness than in the previous million years of inner unity. Because it feels lost, bewildered, unsure of itself, the conscious ego searches obsessively for meaning. The present book has charted the course of its most ambitious voyage of discovery.

And, oddly enough, this voyage has ended very close to its starting point. Nearly three centuries ago, Bishop George Berkeley produced a remarkable anticipation of Einsteinian relativity; he said that motion must always be measured in relation to some other body. Newton had said the same thing; but he believed there was something special about acceleration or motion in a curve – which, he said, could be observed on its own. For example, if you took a bucket of water, placed it on a gramophone turntable, then spun the turntable, the surface of the water would curve. If you could somehow rotate the whole room around the stationary bucket, the surface would remain flat. This, Newton would say, shows whether it is the bucket or the room that is revolving. Berkeley would disagree. The reason the water curves, he would say, is because it is attracted by the whole of the rest of the universe – all the stars and galaxies. Therefore, to make the test fair, you would have to revolve the whole of the rest of the universe around the bucket – and its surface would curve. In a completely empty universe, there would be no centrifugal force.

In 1872, the Austrian physicist Ernst Mach rediscovered this idea, asserting that instead of saying 'absolute space' we should say, 'the whole of the matter in the universe'. That is to say, the totality of the universe exerts force on everything in it – which is even more than the astrologers would claim. The idea was in due course incorporated into the theory of relativity. Which leaves us with the amusing and paradoxical thought that modern physics finds nothing strange in a notion which would have been taken for granted by the architect of the Great Pyramid.

But then he would probably have preferred to express it in the formula of the legendary Hermes Trismegistos: 'As above, so below.'

NOTES

1 The Great Stone Observatories

1 John H. Ingram, *The Works of Edgar Allan Poe*, 4 vols. (London, 1880).
2 Ibid, Memoir, p. 80.
3 Ibid.
4 Daniel Hoffman in *PoPoPoPoPoPoPo*, Robson Books (London, 1973).
5 *Eureka*, also included in *The Science Fiction of Edgar Allan Poe*, ed. Harold Beaver, Penguin Books (1976).
6 Robert Graves, 'The Abominable Mr Gunn', *Collected Short Stories*, Cassell (1968).
7 Robert Graves, *Five Pens in Hand*, Doubleday (1958), p. 55.
8 R. J. C. Atkinson, *Stonehenge*, Penguin Books (1960), p. 168.
9 Jacquetta Hawkes, *The Atlas of Early Man*, Macmillan (1976), p. 22.
10 Arnold L. Lieber, *The Lunar Effect*, Doubleday (New York, 1978).
11 John Gribbin, *The Climatic Threat. What's wrong with our weather?* Fontana/Collins (1978).
12 John Gribbin and Stephen Plageman, *The Jupiter Effect*, Macmillan (1974).
13 Leonard J. Ravitz, 'Periodic Changes in Electromagnetic Fields', *Annals of the New York Academy of Science*, lcviii (1960), cited in Michel Gauquelin, *The Cosmic Clocks*, p. 165.
14 Michel Gauquelin, *The Cosmic Clocks. From Astrology to a Modern Science*, Henry Regnery Company (1967).
15 *The Daily Telegraph*, 20 September 1979.
16 Charles R. Carrington and David Gubbins, 'The Source of the Earth's Magnetic Field', *Scientific American* (February 1979).
17 Euan MacKie, *Science and Society in Prehistoric Britain*, Elek Books (1977).

2 The Pyramid Mystery

1 *Some Trust in Chariots*, ed. Barry Thiering and Edgar Castle, Popular Library (New York, 1972).
2 See Martin Gardner, *Fads and Fallacies in the Name of Science*, Dover Publications (1957), ch. 15.
3 J. Norman Lockyer, *The Dawn of Astronomy* (London, 1894).
4 André Tomas, *Atlantis. From Legend to Discovery*, Sphere Books (1972).
5 Colin Wilson, *The Occult*, Hodder and Stoughton (London, 1971), p. 188.
6 Erich Kahler, *Man the Measure. A New Approach to History*, Pantheon Books (New York, 1943), p. 82.
7 Quoted in I. M. Stewart, *Ecstatic Religion*, Pelican Books (1971), p. 37.
8 Gertrude Levy, *The Gate of Horn*, Faber and Faber (1948), p. 3.
9 R. A. Parker, in *The Place of Astronomy in the Ancient World*, ed. F. R. Hodson, Oxford University Press (London, 1974).
10 Arthur M. Young, *The Bell Notes*, Delacorte Press (New York, 1979), p. 144.

3 The Age of Abstraction

1 See Lynn Thorndike, *History of Magic and Experimental Science*, Columbia University Press (1923–58), vol. 1, p. 219.
2 J. L. E. Dreyer, *A History of Astronomy from Thales to Kepler*, Dover Publications (New York, 1953).
3 Ibid, p. 138.
4 See N. K. Sandars, *The Sea Peoples*, Thames and Hudson (1978).
5 *The Confessions of St Augustine*, transl. F. J. Sheed (London, 1944), p. 197.

4 The Harmony of the World

1 Cited in J. L. E. Dreyer, *Tycho Brahe* (Edinburgh, 1890). Also Dover Publications (New York, 1963).
2 Ibid, p. 18.
3 *De Stella Nova*, ch. 28, quoted in Arthur Koestler, *The Sleepwalkers*, Hutchinson (1959), p. 243.
4 Kepler's 'Diary of Observations', quoted in Dreyer, *Tycho Brahe*.
5 Introduction to *The Harmony of the World*.

6 Cited by Arthur Koestler, *The Sleepwalkers*, p. 243.
7 *The Harmony of the World*.

5 The Lawgivers

1 Quoted in Colin Ronan, *Their Majesties' Astronomers*, The Bodley Head (1967), p. 30.
2 *The Divine Comedy*, Paradiso, Canto 2, lines 31–4.
3 Boris Kouznetsov, *Galilée*, Les Editions Mir (URSS, 1973).
4 Koestler, *The Sleepwalkers*, p. 368.
5 Koestler, p. 453.
6 Jacob Bronowski, *The Ascent of Man*, BBC Publications (1973), p. 208.
7 Newton's notebooks, quoted in Robert S. Richardson, *The Star Lovers*, Macmillan (New York, 1967), p. 27.
8 E. T. Bell, *Man and Mathematics*, Scientific Bookclub (London), p. 114.

6 The Explorers

1 Suetonius, *Lives of the Twelve Caesars*.
2 Cited in *Hutchinson's Splendour of the Heavens*, ed. T. E. R. Phillips and W. H. Steavenson, p. 389.
3 Willey Ley, *Watchers of the Skies*, Viking Press (1963), p. 443.
4 *Asimov's Biographical Encyclopedia of Science and Technology*, Pan Books (1972).

7 A Voyage Round the Planets

1 John Glenn, 1969.
2 Isaac Asimov, *The Tragedy of the Moon*, Coronet Books (London, 1972).
3 Cited in Arthur Rosenblum, *Unpopular Science*, Running Press (Philadelphia, 1974), p. 19.
4 See, for example, H. J. Eysenck, 'Astrology – Science or Superstition', *Encounter* (December 1979).
5 Published in *Philosophical Transactions*, 1781 and 1784.
6 William Lilly, *An Introduction to Astrology*,

reprinted by Newcastle Publishing Co. Inc. (Hollywood, California, 1972), p. 35.

8 The World Turned Inside Out

1 Edmund Taylor Whittaker, *History of the Theories of Aether and Electricity*, vol. 2, Thomas Nelson and Sons (1900–1926).
2 Quoted in Whittaker.
3 Whittaker.
4 Quoted in Whittaker, *From Euclid to Eddington*, Dover Publications (New York, 1958), p. 112.

9 In the Beginning

1 See Thomas R. Geballe, 'The Central Parsec of the Galaxy', *Scientific American* (July 1979).
2 Lloyd Motz, *The Universe. Its Beginning and End*, Abacus (1977).
3 Fred Hoyle, *Ten Faces of the Universe*, Heinemann Educational (1977), p. 159.
4 Quoted in Nigel Calder, *The Violent Universe*, BBC Publications (1969), p. 149.
5 E. A. Milne, 'The Fundamental Concepts of Natural Philosophy', in *Theories of the Universe. From Babylonian Myth to Modern Science*, ed. Milton K. Munitz, The Free Press (New York, 1957), p. 354.
6 Milne, *Relativity, Gravitation and World Structure*, Oxford University Press (1935), p. 3.
7 Martin Gardner undertakes to answer this objection in *Relativity for the Million*, Cardinal, Pocket Books (1962), ch. 8 ('The Twin Paradox'), explaining that 'the stay-at-home does not move relative to the universe'. The objection was first raised by Professor Herbert Dingle.

Postscript

1 *Bestimmung des Menschen*, Book I.
2 David Foster, *The Intelligent Universe. A Cybernetic Philosophy*, G. P. Putnam's Sons (New York, 1975).

BIBLIOGRAPHY

ABETTI, Giorgio, *The Exploration of the Universe*, Faber and Faber, London, 1965.

ASIMOV, Isaac, *Realm of Algebra*, Fawcett Publications Inc., USA, 1961.
The World of Carbon, Collier Books, USA, 1962.
The Kingdom of the Sun, Collier Books, USA, 1962.
The Tragedy of the Moon, Mercury Press Inc., London, 1972.

ATKINSON, R. J. C., *Stonehenge*, Penguin Books, Middlesex, 1960.

BORN, Max, *Einstein's Theory of Relativity*, Dover Publications Inc., New York, 1962.

BRONOWSKI, J., *The Ascent of Man*, British Broadcasting Corporation, London, 1973.

BROWN, Hanbury, FRS, *Man and the Stars*, Oxford University Press, Oxford, 1978.

CLARK, Ronald W., *Einstein, The Life and Times*, World Publishing Company, New York, 1971.

D'ABRO, A., *The Rise of the New Physics*, vol. 1, Dover Publications Inc., USA, 1951.

DREYER, J. L. E., *A History of Astronomy from Thales to Kepler*, Dover Publications Inc., USA, 1953.
Tycho Brahe, Dover Publications Inc., USA, 1963.

EDDINGTON, Sir Arthur, MA, DSc, LLD, FRS, *The Expanding Universe*, Penguin Books, Middlesex, 1940.
The Internal Constitution of the Stars, Dover Publications Inc., New York, 1959.

EDWARDS, I. E. S., *The Pyramids of Egypt*, Penguin Books, Middlesex, 1947.

EINSTEIN, Albert, with LORENTZ, H. A., MINKOWSKI, H., and WEYL, H., *The Principle of Relativity. A Collection of Original Memoirs on the Special and General Theory of Relativity*, Methuen & Co., London, 1923.

EINSTEIN, Albert, *Relativity. The special and the general theory*, Methuen & Co., London, 1946.
The Evolution of Physics, Cambridge University Press, London, 1938.

FIRSOFF, V. A., *Strange World of The Moon*, Science Editions Inc., New York, 1962.

FRENCH, A. P., *Einstein. A Centenary Volume*, Heinemann Educational Books, London, 1979.

FRENCH, Bevan M., *The Moon Book*, Penguin Books, Middlesex, 1977.

FREUNDLICH, Erwin, *The Foundations of Einstein's Theory of Gravitation*, Methuen & Co., London, 1924.

GAMOW, George, *The Creation of the Universe*, The New American Library, New York, 1957.
Thirty Years That Shook Physics, Doubleday & Co. Inc., New York, 1966.

GARDNER, Martin, *Relativity for the Million*, Pocket Books Inc., New York, 1965.

GILZIN, Karl, *Travel to Distant Worlds*, Foreign Languages Publishing House, Moscow, 1957.

GRIBBIN, John, *White Holes. Cosmic Gushers in the Universe*, Paladin, St Albans, Herts., 1977.

HODSON, F. R., *The Place of Astronomy in the Ancient World. A joint symposium of The Royal Society and The British Academy*, Oxford University Press, London, 1974.

HOFFMANN, Banesh, *The Strange Story of the Quantum*, Dover Publications Inc., New York, 1959.
Albert Einstein, Granada Publishing, St Albans, Herts., 1975.

HOYLE, Fred, FRS, *Frontiers of Astronomy*, Signet Science Library Books, New York, 1963.

Ten Faces of the Universe, Heinemann Educational Books, London, 1977.

HOYLE, Fred, and WICKRAMASINGHE, Chandra, *Lifecloud. The origin of life in the universe*, J. M. Dent & Sons, London, 1978.

JEANS, Sir James, MA, DSc, LLD, FRS, *The Universe Around Us*, Cambridge at the University Press, London, 1929.
The Mysterious Universe, Cambridge at the University Press, London, 1930.
The Stars in their Courses, Cambridge at the University Press, London, 1931; Pelican Books, London, 1939.

KOESTLER, Arthur, *The Sleepwalkers. A History of Man's Changing Vision of the Universe*, Hutchinson, London, 1959.

KOUZNETSOV, B., *Galilée*, Les Editions Mir, URSS, 1973.

LANDAU, L. D., and LIFSHITZ, E. M., *Quantum Mechanics. Non-Relativistic Theory*, Pergamon Press, London, 1958; Addison-Wesley Publications, USA.

LEY, Willy, *Ranger to the Moon*, Signet Science Library Books, New York, 1965.
Watchers of the Skies, The Viking Press Inc., New York, 1963.

MILNE, E. A., MA, DSc, FRS, *Relativity, Gravitation and World-Structure*, Oxford University Press, London, 1935.

MITTON, Simon, MA, PhD, *The Cambridge Encyclopaedia of Astronomy*, Jonathan Cape, London, 1979.

MOORE, Patrick, OBE, DSc (Hon), FRAS, *The Atlas of the Universe*, Mitchell Beazley and George Philip & Son, London, 1970.

MOORE, Patrick, and NICOLSON, Iain, *Black Holes in Space*, Ocean Books, London, 1974.

MOORE, Patrick, *Guide to Comets*, Lutterworth Press, London, 1977.

MOSZKOWSKI, Alexander, *Einstein. The Searcher*, Methuen & Co., London, 1921.

MOTZ, Lloyd, *The Universe. Its Beginning and End*, Sphere Books, London, 1977.

MURCHIE, Guy, *Music of the Spheres*, vol. 1, Dover Publications Inc., New York, 1967.

POINCARÉ, Henri, *The Value of Science*, Dover Publications Inc., New York, 1958.

RICE, James, MA, *Relativity. An Exposition without Mathematics*, Benn's Sixpenny Library, London, 1928.

RICHARDSON, Robert S., *The Star Lovers*, The Macmillan Company, New York, 1967.

RONAN, Colin A., *Their Majesties' Astronomers*, The Bodley Head, London, 1967.

ROTHMAN, Milton A., *The Laws of Physics*, Penguin Books, Middlesex, 1966.

SADIL, Josef, *The Moon and the Planets*, Paul Hamlyn, London, 1963.

SANDNER, Werner, *Satellites of the Solar System*, Faber and Faber, London, 1965.

SIMON, Tony, *The Search for Planet X*, Scholastic Book Services, New York, 1969.

SKILLING, William T. and RICHARDSON, Robert S., *Astronomy*, Henry Holt & Co. Inc., New York, 1951.

SMITH, F. Graham, *Radio Astronomy*, Penguin Books, Middlesex, 1962.

STAAL, Julius D., FRAS, *Focus on Stars. Everyman's Guide to Astronomy*, Newnes, London, 1963.

TAYLOR, J. G., *Quantum Mechanics: An Introduction*, George Allen and Unwin, London, 1970.
Black Holes: The End of the Universe?, Souvenir Press, London, 1973.
New Worlds in Physics, Faber and Faber, London, 1974.

WILKINS, H. P., and MOORE, Patrick, *The Moon*, Faber and Faber, London, 1961.

INDEX

Numbers in italics refer to illustrations

ACKNOWLEDGEMENTS

The author and publishers wish to thank the following for permission to reproduce photographs:

Barnaby's Picture Library, page 19 (below); BBC Hulton Picture Library, pages 51, 56, 140 and 146; Bettman Archive, page 195; Biblioteca Estense, Modena, pages 187, 192, 193, 202, 206 and 207; Bibliothèque Nationale, pages 134/135; Camera Press, pages 179 and 244; J. Allan Cash, pages 55 (below), 163 and 185; Cavendish Laboratory, University of Cambridge, pages 229 and 230; Department of Navigation, Hydrography and Oceanography, Istanbul, page 39; Mary Evans Picture Library, title page, pages 11, 15, 19 (above), 22, 25, 26, 33, 41, 43, 52, 58, 87, 114, 122, 136, 138 (right), 151, 171, 172, 178, 180 (above and below), 181, 189 (above and below), 208 and 232; Fortean Picture Library, page 31; Fotomas Index, pages 80, 93 (above and below), 96/97, 113 and 177; Photographie Giraudon, pages 45 and 64; Fay Godwin, page 16 (above); Robert Harding Associates, page 55 (above); John Hillelson Agency, pages 28/29 (photo Dr Georg Gerster), 88, 89, 109 and 225 (photos Erich Lessing); Michael Holford, pages 8/9 detail of *Stonehenge in a Storm* by J. M. W. Turner, 16 (below), 21, 48, 53, 60, 158 (above) and 161; John Holmes, pages 7, 13, 85 and 175; Lowell Observatory Photograph, page 198 (above and below); Mansell Collection, pages 69, 77, 82, 83, 90, 95, 102, 103, 121, 156, 158 (below left), 165; Marconi Company Ltd, page 217; Marshall Cavendish, pages 34/35 (illustration by David Roberts); MGM, page 223 (above and below); Musée d'Histoire de l'Education, page 149; National Maritime Museum, London, pages 37 and 154/155; Popperfoto, pages 216, 218, 227 and 239; Ann Ronan Picture Library, pages 23, 44, 67, 68, 98, 106, 110, 117, 126, 138 (left), 139 (left and right), 144 and 158 (below right); Royal Photographic Society, page 166; Scala, pages 62, 73, 78, 101, 104/105, 120, 123, 125, 128, 131, 153 and 254; Science Photo Library, pages 184, 204, 212, 237, 240, 242/243 and 247; Ronald Sheridan, pages 20 and 72; Space Frontiers, pages 196 (left and right), 197 and 200/201; Staatliche Museum, East Berlin, page 74; The Tate Gallery, London, pages 142/143 and 214; Western Americana, page 182.